Vectorial
Analysis

Vectorial Analysis

PARABOLOID

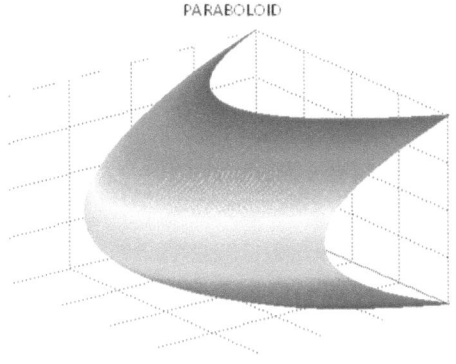

SIMPLIFIED MATH

Ing. Mario Castillo

Número de Control de la Biblioteca del Congreso de EE. UU.:   2012908941
ISBN: Tapa Dura   978-1-4633-2978-5
               Tapa Blanda   978-1-4633-2977-8
               Libro Electrónico   978-1-4633-2976-1

Para pedidos de copias adicionales de este libro, por favor contacte con:
Palibrio
1663 Liberty Drive
Suite 200
Bloomington, IN 47403
Llamadas desde los EE.UU. 877.407.5847
Llamadas internacionales +1.812.671.9757
Fax: +1.812.355.1576
ventas@palibrio.com
408317

INDICE

PROLOGUE

THE SOLUTION FOR THE PROBLEMS PRESENTED IN THIS BOOK ARE SOLVED WITH ALGEBRA, ANALYTIC GEOMETRY, DIFFERENTIAL AND INTEGRAL CALCULUS GEOMETRY, MATLAB AND VECTOR ANALYSIS.

CHAPTER I ARE VECTOR AND SCALARS, WE SOLVE PROBLEMS OF PHYSICS (FORCES) AND DYNAMICS.

CHAPTER II DOT AND DOT PRODUCT, WE SOLVE PROBLEMS OF ADD, SUBTRACT, MULTIPLY AND VECTOR DIVISION IN THREE DIMENSIONS, ANGLES AND PLANES.

CHAPTER III DERIVATION OF VECTORS AND SCALARS, VELOCITY AND ACCELERATION AND ITS DIAGRAMS (VISUALIZATION). IN THREE DIMENSIONS.

CHAPTER IV GRADIENT OF ADD, SUBTRACT MULTIPLY AND DIVIDE FUNCTIONS. GRAPHICS IN TRIDIMENSIONAL FUNCTIONS, PERPENDICULAR VECTOR AT A SURFACE. ONE FUNCTION.

CHAPTER V. VECTOR INTEGRATION PROBLEMS OF VELOCITY AND ACCELERATION OF PARTICLES IN ONE PATH, FORCE BY DISTANCE. UNIT VECTORS.

CHAPTER VI. TRANSFORMATION OF COORDINATES SPHERICAL, CYLINDRICAL, PARABOLIC, ELLIPTIC UNIT VECTORS, VECTORS, ORTHOGONAL FUNCTIONS, THREE DIMENSION GRAPHICS.

THE AUTHOR OF THIS BOOK THANKS AT THE INSTRUCTORS OF THE ELECTRICAL ENGINEERING DEPARTMENT THE UNIVERSITY OF TEXAS PAN AMERICAN OF EDINBURG, DR HEINRICH FOLTZ AND DR BEN GHALIA MOUNIR BY ALL ITS ADVISES.

CHAPTER I

VECTOR AND SCALARS

1   WHICH OF THE FOLLOWING ARE SCALARS AND WHICH ARE VECTORS?

(A)	KINETIC ENERGY	SCALAR
(B)	ELECTRIC FIELD INTENSITY	VECTOR
(C)	ENTROPY	SCALAR
(D)	WORK	SCALAR
(E)	CENTRIFUGAL FORCE	VECTOR
(F)	TEMPERATURE	SCALAR
(G)	GRAVITATIONAL POTENTIAL	SCALAR
(H)	CHARGE	SCALAR
(I)	CHEARING STRESS	VECTOR
(J)	FRECUENCY	SCALAR

2   AN AIRPLANE TRAVELS 200 MILES DUE WEST AND THEN 150 MILES 60 DEGREES NORTH OF WEST. DETERMINE THE RESULTANTE DISPLACEMENT.

R1 = 200;

ANG1 = pi;

R2 = 150;

ANG2 = 2*pi/3;

```
Rx = (R1*cos (ANG1) + R2*cos (ANG2));
Ry = R1*sin (ANG1) + R2*sin (ANG2);
R = (Rx^2 + Ry^2) ^.5;
fprintf ('R %f /n', R);
fprintf ('Rx %f /n', Rx);
fprintf ('Ry %f /n', Ry);
theta = (180/pi)*(atan(Ry/Rx));
fprintf('theta %f /n', theta);
RESULT R = 304.138127 /n Rx= -275.000000 /n Ry = 129.903811 /n
Theta = -25.284996 /n
```

3 FIND THE RESULTANT OF THE FOLLOWING DISPLACEMENTS: A. 20 MILES 30° SOUTH OF EAST; B. 50 MILES DUE WEST; C. 40 MILES NORTHEAST; D. 30 MILES 60 DEGREES SOUTH H OF WEST.

```
R1 = 20;
ANG1 = - pi/6;
R2 = 50;
ANG2 = - pi;
R3 = 40;
ANG3 = pi/4;
R4 = 30;
ANG4 = - 2*pi/3;
Rx = R1*cos(ANG1) + R2*cos(ANG2) + R3*cos(ANG3) + R4*cos(ANG4);
Ry = R1*sin(ANG1) + R2*sin(ANG2) + R3*sin(ANG3) + R4*sin(ANG4);
R = ( Rx^2 + Ry^2)^.5;
fprintf('R %f /n', R);
fprintf('Rx %f /n', Rx);
fprintf('Ry %f /n', Ry);
theta = (180/pi)*(atan(Ry/Rx));
SOLUTION R 20.866494 /nRx -19.395221 /nRy -7.696491 /n
```

4   SHOW GRAPHICALLY THAT - (A - B) = - A + B

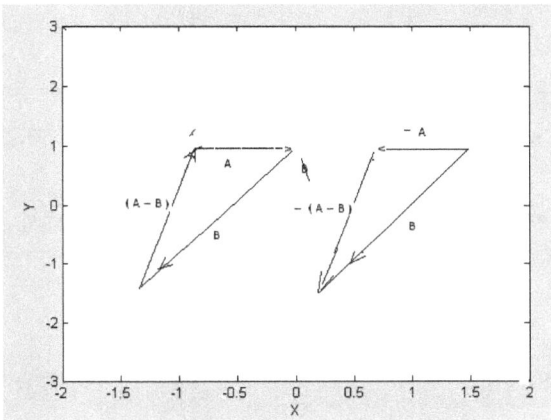

5   F ABCDEF ARE THE VERTICES OF A REGULAR HEXAGON, FIND THE RESULTANT OF THE FORCES REPRESENTED BY THE VECTORS AB, AC, AD, AE AND AF.

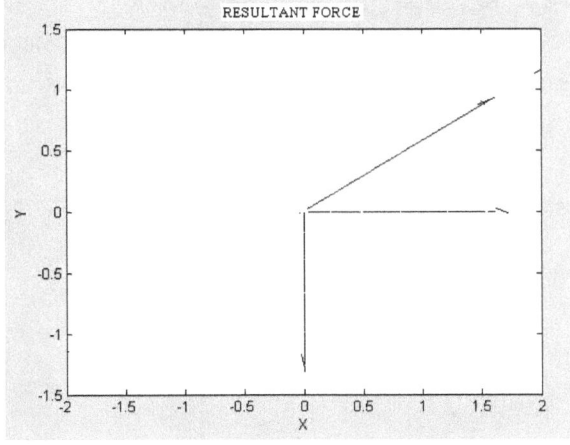

$Fx = 150 + 200(\cos 30) = 150 + 173.2 = 323.2lb$

$Fy = 100 - 200(\sin (30)) = 100 - 100 = 0 \ lb$

6   IF A AND B ARE GIVEN VECTORS SHOW THAT (A) | A + B | <= | A | + B

(B) | A - B | < = | A | - | B |

= Ax + Ay,   B = Bx + By

|A + B| = |Ax + Bx + Ay + By| <= | Ax + Ay| + | Bx + By| ARE EQUALS IF THE FORCES HAVE EQUAL SYMBOLS, IF HAVE OPPOSITE SIGNS ARE MINORS.

|A – B| >= | A| - |B | => |Ax +Ay - Bx - By| = |Ax + Bx| - |Ay + By|

ARE EQUALS IF THE FORCES HAVE EQUAL SYMBOLS, IF THE FORCES BOTH HAVE DIFFERENT SYMBOLS ARE BIGGER.

7.   DEMOSTRATE THAT (A) | A + B + C | <= | A | + | B | + | C |

A = Ax + Ay,   B = Bx + By,   C = Cx + Cy

|A + B + C| = |Ax + Bx + Ay + By + Cx + Cy| <= |Ax + Ay| + |Bx + By| + |Cx + Cy| ARE EQUALS IF THE FORCES HAVE EQUAL SYMBOLS, IF HAVE OPPOSITE SIGNS ARE MINORS.

8.   ONE MEN TRAVELING IN SOUTH DIRECTION AT 15 MI/HR OBSERVE THAT THE WIND COMES FROM THE WEST. INCREASING HIS VELOCITY AT 25 MI/HR THE WIND COMES FROM SOUTH EAST. FIND THE DIRECTION AND VELOCITY OF THE WIND.

X = (15^2 + 10^2) ^.5 = 18

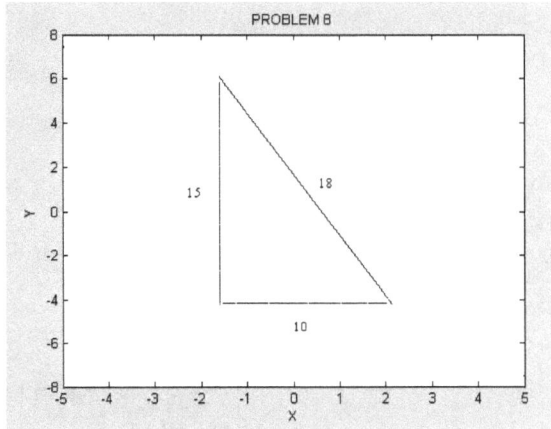

9   100 LB WEIGHT IS SUSPENDED FROM THE CENTER OF A ROPE AS
     SHOWN IN THE ADJOINING FIGURE. DETERMINE THE TENSION T IN
     THE ROPE.

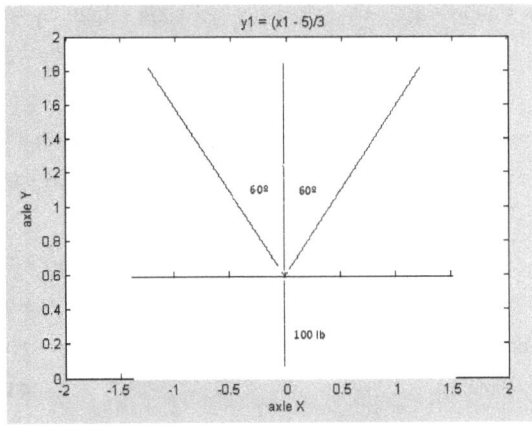

R1 = 100;

ANG1 = pi/6;

R2 = 100;

ANG2 = pi/6;

Ry = R1*sin (ANG1) + R2*sin (ANG2);

R = Ry;

fprintf('R %f /n', R);

fprintf('Ry %f /n', Ry);

SOLUTION   R 100.000000 /nRy 100.000000 /n

---

10   SIMPLIFY   2A + B + 3C - { A - 2B - 2 (2A - 3B - C)}

2A + B + 3C - { A - 2B - 4A + 6B + 2C } = 2A + B + 3C - A + 2B + 4A -6B - 2C

$$= 5A - 3B + C$$

---

11   IF a AND b ARE NON-COLINEAR VECTORS AND A = (x + 4y) a + (2x + y + 1) b AND B

= (y − 2x + 2)a + (2x − 3y -1) b FIND x AND y SUCH THAT 3A = 2B .

3 (x + 4y)a + 3(2x + y +1)b = 2(y - 2x + 2)a + 2(2x -3y − 1)b

(3x + 12y)a + (6x + 3y + 3)b = (2y − 4x + 4)a + (4x - 6y − 2)b

We have two equations vector "a" and vector "b"

(3x + 12y) = (2y − 4x + 4)   AND   (6x + 3y +3) = (4x − 6y -2)

7x + 10y = 4 (A)   2x + 9y = -5 (B)

x = (4 - 10y)/7 (C)

(C) AND (B)   2 (4 -10y)/7 + 9y = -5

(8 − 20y)/7 + 9y = -5

Multiply both members by 7   (8 − 20y) + 63y = - 35

Simplify   43y = - 43

Y = - 1

Replacing "y " in (A) or (B)   2x + 9y = -5

2x - 9 = -5

2x = 4

X = 2

12 IF a, b, c ARE NON-COPLANAR VECTORS DETERMINE WHETER THE VECTORS r1 = 2a -3b + c, r2 = 3a - 5b + 2c AND r3 = 4a – 5b + c ARE LINEARLY INDEPENDENT OR DEPENDENT.

$$\text{DET} \begin{vmatrix} 2 & -3 & 1 \\ 3 & -5 & 2 \\ 4 & -5 & 1 \end{vmatrix} = -10 - 15 - 24 + 20 + 20 + 9 = 0 \quad \text{THE DET} = 0$$

IS LINERLY DEPENDENT.

13. IF a, b, c ARE VECTORS NON-COPLANAR DETERMINE IF THE VECTORS r1 = 2a -3b + c, r2 = 3a - 5b + 2c Y r3 = 4a – 5b + c ARE LINEARLY INDEPENDENT.

DEPENDIENTE.

$$\text{DET} \begin{vmatrix} 2 & -3 & 1 \\ 3 & -5 & 2 \\ 4 & -5 & 1 \end{vmatrix} = -10 - 15 - 24 + 20 + 20 + 9 = 0 \quad \text{THE DET} = 0$$

IS LINERLY DEPENDENT.

14 IF A AND B ARE GIVEN VECTORS REPRESENTING THE DIAGONALS OF A PARALLELOGRAM, CONSTRUCT THE PARALLELOGRAM.

AB = A/2 + B/2 = (A + B)/2

C = A + B = 2 (AB) = A + B

15 (a) IF O IS ANY POINT WITHIN TRIANGLE ABC AND P, Q, R ARE MID POINTS OF THE SIDES AB, BC, CA RESPECTIVELY, PROVE THAT OA + OB + OC = OP + OQ + OR.

OA + OB = AB   OP + OR = AB/2

OB + OC = BC   OR + OQ = CB/2

OC + OA = CA   OQ + OP = CA/2

ADDING AND SIMPLIFYING

2 OA + 2 OB + 2 OC = AB + BC + CA, 2 OP + 2OR + 2OQ = (AB + CB + CA)/2

OA + OB + OC = (AB `+ BC + CA)/2, OP + OR + OQ = (AB + CB + CA)/4

---

16   IN THE ADJOINING FIGURE, ABCD IS A PARALLELOGRAM WITH P AND Q THE MIDPOINTS OF THE SIDES BC AND CD RESPECTIVELY. PROVE THAT AP AND AQ TRISECT DIAGONAL BD AT THE POINTS E AND F.

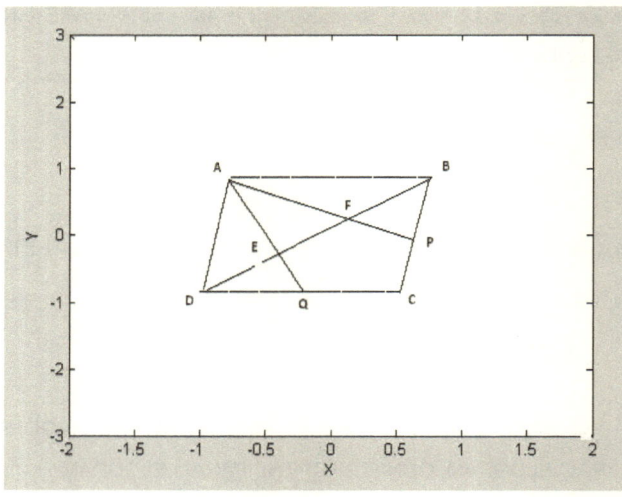

DE = EF = FB = a + b,   EF = EQ + a/2 + b/2 + FP

DE = (a + b)/3;   EF = b/3 - a/6 + a/2 + b/2 + b/6

DE = a/2 + EQ = (a + b)/3   - a/3

EQ = (a + b)/3 - a/2 = b/3 - a/6;   EF = b/3 - a/3 + a/2 + b/2

FB = b/2 - FP = (a + b)/3   EF = 5b/6 + a/6 = (a + b) /3

FP = b/2 - (a + b)/3   - 2a/3 + a/2 = b/3 -5b/6

FP = b/6 - a/3;   - a/6 = - b/2

$$a = 3b$$
$$EF = b/3 + 2b/6 + a/6$$
$$EF = b/3 + a/9 + a/6$$

17 DEMOSTRATE THAT THE MEDIANS OF ONE TRIANGLE JOINING IN ONE COMMON WHICH IS THE TRISECTION POINT OF THE MEDIANS?

M1 = 2 a + c = 2 b + c   M2 = 2 c + a = 2 b + a   M2 = 2 c + a = 2 b + a

a = b   c = b   a = c

a = c = b

M1 = 2 a + a = 3 a   M2 = 2 b + b = 3 b   M3 = 2 c + c = 3 c

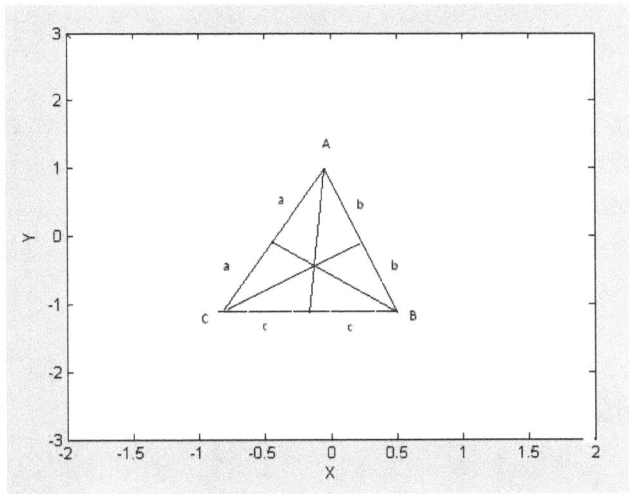

18 PROVE THAT THE BISECTORS ANGLE OF ONE TRIANGLE JOINING IN ONE COMMON POINT.

$c^2 = r^2 + s1^2$ $a^2 = r^2 + s2^2$ $b^2 = r^2 + s3^2$

$r^2 = c^2 - s1^2$ $r^2 = a^2 - s2^2$ $r^2 = b^2 - s3^2$

$c^2 - s1^2$ = $a^2 - s2^2$ = $b^2 - s3^3$

IS ONE CIRCLE OF RADIO r.

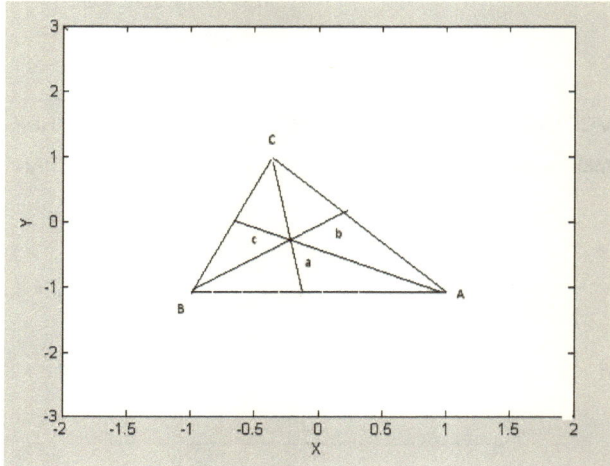

---

19 LET THE POSITION VECTORS OF POINTS P AND Q RELATIVE TO AN
    ORIGIN O BE GIVEN BY p AND q RESPECTIVELY. IF R IS A POINT WHICH
    DIVIDES LINE PQ INTO SEGMENTS WHICH ARE IN THE RATIO m:n
    SHOW THAT THE POSITION VECTOR OF R IS GIVEN BY r = (mp + nq)/(m
    + n) AND THAT THIS IS INDEPENDENTN OF THE ORIGIN.

---

mp + nq = r (m + n)

r = (mp + nq)/(m + n)

---

20 IF r1, r2, r3 ....rn ARE THE POSITION VECTORS OF MASES m1,
    m2, m3 ....mn RESPECTIVELY RELATIVE TO AN ORIGIN O.
    SHOW THAT THE POSITION VECTOR OF THE CENTROID IS GIVEN BY

---

r = (r1m1 + r2m2 + ...rnmn)/(m1 + m2 + ...mn)

AND THAT THIS IS INDEPENDENT OF THE ORIGIN.

m1r1 + m2r2 + ...mnrn = r (m1 + m2 + ...mn)

r = (m1r1 + m2r2 + ...mnrn)/(m1 + m2 + ...mn)

21 A QUADRILATERAL ABCD HAS MASSES OF 1, 2, 3, AND 4 UNITS
    LOCATED RESPECTIVELY AT ITS VERTICES A (-1, -2, 2), B (3, 2, -1), C (1,
    -2, 4) AND D (3, 1, 2) FIND THE COORDINATES OF THE CENTROID.

(1) (-1) + (2) (3) + (3) (1) + (4) (3) = x (1 + 2 + 3 + 4) = 10x

- 1 + 6 + 3 + 12 = 10x

    20 = 10 x

      X = 2

(1) (-2) + (2) (2) + (3) (-2) + (4) (1) = 10 Y

-2 + 4 -6 + 4 = 10 Y

       Y = 0

(1) (2) + (2) (-1) + (3) (4) + (4) (2) = 10 Z

   2 - 2 + 12 + 8 = 10 Z

       Z = 2

---

22 SHOW THAT THE EQUATION OF A PLANE WHICH PASSES THROUGH
    THREE GIVEN POINTS A, B, C NOT IN THE SAME STRAIGHT LINE AND
    HAVING POSITION VECTORS

a, b, c RELATIVE TO AN ORIGIN O, CAN BE WRITTEN r = (ma + nb + pc)/
(m + n + p) WHERE m, n, p ARE SCALARS. VERIFY THAT THE EQUATION
IS INDEPENDENT OF THE ORIGIN.

    ma + nb + pc = r(m + n + p)

       r = (ma + nb + pc)/(m + n + p)

---

23 THE POSITION VECTORS OF POINTS P AND Q ARE GIVEN BY r1 = 2i +
    3j - k, r2 = 4i -3j +2k.

DETERMINE PQ IN TERMS OF i, j, k AND FIND ITS MAGNITUDE.

PQ = (x2 - x1)I + (y2 - y1)j + (z2 - z1)k

PQ = (4 - 2)i + (-3 - 3)j + (2 - (-1))k

PQ = 2i - 6j + 3k | PQ | = (2*2 + (-6)*(-6) + (3)*(3)) ^(.5)

                      | PQ | = (4 + 36 + 9)^:5

                      | PQ | = 7

---

24   IF A = 3i - j - 4k B = -2i + 4j -3k C = I + 2j - k, FIND (A) 2A - B + 3C, (B) | A + B
     + C |, (C) | 3A - 2B + 4C | (D) A UNIT VECTOR PARALLEL TO 3A - 2B + 4C.

A = [3 -1 -4];

B = [-2 4 -3];

C = [1 2 -1];

X1 = 2*A - B + 3*C;

X2 = A + B + C

X3 = 3*A - 2*B + 4*C

X3 = ((X3(1,1))^2 + (X3(1,2))^2 + (X3(1,3))^2)^.5;

X2 = ((X2(1,1))^2 + (X2(1,2))^2 + (X2(1,3))^2)^.5;

fprintf(' X1 %f /n', X1);

fprintf(' X2 %f /n', X2);

fprintf(' X3 %f /n', X3);

---

25   THE FOLLOWING FORCES ACT ON A PARTICULE P: F1 = 2i + 3j - 5k F2
     = -5i + j + 3k, F3 = i - 2j + 4k, F4 = 4i - 3j - 2k, MEASURED IN POUNDS.
     FIND (a) THE RESULTANT OF THE FORCES (b) THE MAGNITUDE OF
     THE RESULTANT.

R = F1 + F2 + F3 + F4 = (2i + 3j - 5k) + (-5i + j + 3k) + (I - 2j + 4k) +

(4i - 3j - 2k)

R = 2i - j + 0k |R| = (2*2 + (-1)*(-1))^.5 = (5)^:5

---

26   IN EACH CASE DETERMINE WHETHER THE VECTORS ARE LINEARLY
     INDEPENDENT OR LINEARLY DEPENDENT:

(a) A = 2i + j -3k B = I - 4k C = 4i + 3j - k (b) A = I - 3j + 2k B = 2i - 4j - k C

   = 3i + 2j - k

A(3,1) = 4, A(3,2) = 3, A(3,3) = -1;

B(1,1) = 1, B(1,2) = -3, B(1,3) = 2;

B(2,1) = 2, B(2,2) = -4, B(2,3) = -1;

B(3,1) = 3, B(3,2) = 2, B(3,3) = -1;

det (A)

det (B)

ans = 0 LINEARMENTE DEPENDENTE

ans = 41 LINEARMENTE INDEPENDIENTE

---

27. PROVE THAT ANY FOUR VECTORS IN TRI DIMENSIONS MUST BE LINEARLY DEPENDENT.

A1 = A1xI + A1yJ + A1zk, A2 = A2xI + A2yj + A2zk, A3 = A3xi + A3yj + A3zk,

A4 = A4xi + A4yJ + A4zk

$$Det\ A = \begin{vmatrix} A1x & A1y & A1z & 0 \\ A2x & A2y & A2z & 0 \\ A3x & A3y & A3z & 0 \\ A4x & A4y & A4z & 0 \end{vmatrix} = 0 \quad ES\ LINEARMENTE\ DEPENDIENTE$$

---

28. PROVE THAT THE NECESARY AND SUFICIENT CONDITION FOR THAT THE VECTORS a = A1i + A2j + A3k, B = B1i + B2j + b3k, C = C1i + C2j + C3k BE LINEARLY IN DEPENDENT IS THAT THE DETERMINANT

$$DET\ \begin{vmatrix} A1 & A2 & A3 \\ B1 & B2 & B3 \\ C1 & C2 & C3 \end{vmatrix} \quad BE\ DIFFERENT\ OF\ ZERO.$$

DET = A1B2C3 + B1C2A3 + C1B3A2 - C1B2A3 - A1C2B3 - C3B1A2 =
DIFFERENT OF ZERO.

29. PROVE THAT THE VECTORS A = 3i + j - 2k, B = -i + 3j + 4k, C = 4i - 2j - 6k
CAN FORM THE SIDES OF ONE TRIANGLE (b) FIND THE LONGITUDE
OF THE MEDIANS OF THE TRIANGLE.

$(2/3)(14)^{.5} = (56/9)^{.5} = (6)^{.5}$

$(2/3)(26)^{.5} = ((104/9)^{.5} = (11.4)^{.5} = (1/2)(104)^{.5}$

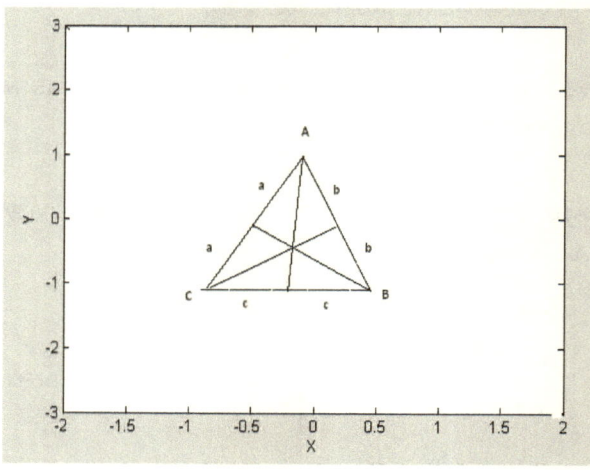

$(2/3)(56)^{.5} = (224/9)^{.5} = (25)^{.5} = (1/2)(225)$

$$A \times B = \begin{vmatrix} I & j & k \\ 3 & 1 & -2 \\ -1 & 3 & 4 \end{vmatrix} = 4i + 9k + 2j - (-k - 6i + 12j) = 10 I + 10k - 10j$$

$$\text{DET} \begin{vmatrix} 2 & -3 & 1 \\ 3 & -5 & 2 \\ 4 & -5 & 1 \end{vmatrix} = -10 - 15 - 24 + 20 + 20 + 9 = 0 \quad \text{THE DET} = 0$$

SON LINERMENTE DEPENDIENTE.

$$\text{BxC} = \begin{vmatrix} I & j & k \\ -1 & 3 & 4 \\ 4 & -2 & -6 \end{vmatrix} = -18i + 2k + 16j - (12k - 8i + 6j) = -10i - 10k + 10j$$

Triangle equilateral M = sin 60 = M1/ (300)^.5, M = (1/2)(300)^.5(3)^.5

(900/4)^.5 = (225)^.5

M1 = 2 a + c = 2 b + c   M2 = 2 c + a = 2 b + a   M3 = 2 a + b = 2 c + b

a = b   c = b   a = c

a = c = b

M1 = 2 a + a = 3 a   M2 = 2 b + b = 3 b   M3 = 2 c + c = 3 c

---

30. GIVEN THE SCALAR FIELD DEFINED BY (Φ X, Y, Z) = 4 Y Z^3 + 3 X Y Z - Z^2 + 2. FIND (A) Φ (1, -1, -2) (B) Φ (0, -3, 1).

X = 1; X1 = 0;

Y = -1; Y1 = -3;

Z = -2; Z1 = 1;

R = 4*Y*(Z^3) + 3*X*Y*Z - Z^2 + 2 ;

R1 = 4*Y1*(Z1^3) + 3*X1*Y1*Z1 - Z1^2 + 2;

fprintf(' R %f /n', R);

fprintf(' R1 %f /n', R1);

R 36.000000 /n R1 -11.000000 /n

---

31. GRAPH THE FIELDS OF VECTORS DEFINED BY

---

a) V(x,y) = Xi - Yj   (b) V(x,y) = Yi - Xj,

(c) $v(x,y,z) = (xi + yj + zk)/(x^2 + y^2 + z^2)^{\wedge}.5$

V(x,y) = Xi – Yj

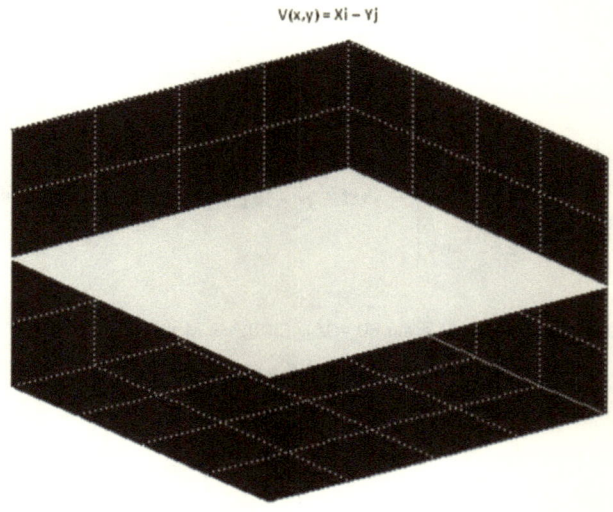

V(x,y) = Yi – Xj,

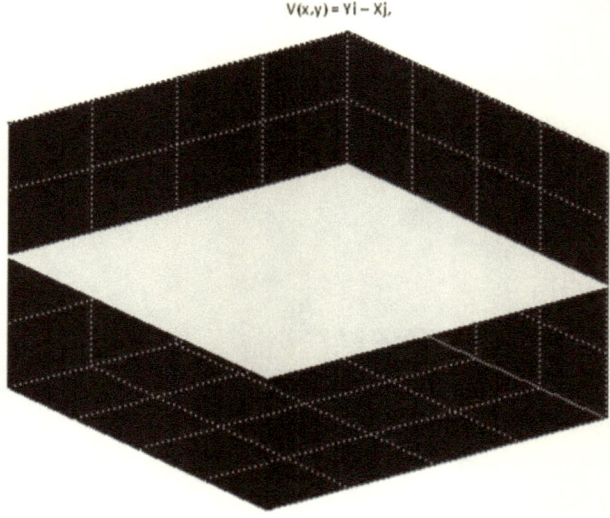

v(x,y,z) = (xi + yj + zk),

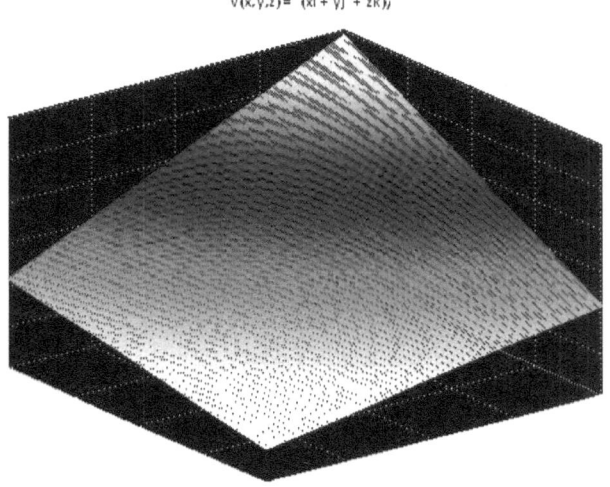

29 GIVEN THE SCALAR FIELD DEFINED BY Φ (X, Y, Z) = 4 Y Z^3 + 3 X Y Z - Z^2 + 2.

FIND (A) Φ (1, -1, -2) (B) Φ (0, -3, 1).

X = 1; X1 = 0;

Y = -1; Y1 = -3;

Z = -2; Z1 = 1;

R = 4*Y*(Z^3) + 3*X*Y*Z - Z^2 + 2 ;

R1 = 4*Y1*(Z1^3) + 3*X1*Y1*Z1 - Z1^2 + 2;

fprintf(' R %f /n', R);

fprintf(' R1 %f /n', R1);

R 36.000000 /n R1 -11.000000 /n

CHAPTER II

THE DOT AND CROSS PRODUCT

---

1   EVALUATE: (a) k.(i + j), (b) (l - 2k).(j + 3k), (c) (2i - j + 3k).(3i + 2j - k)

---

k.(l + j) = 0

(l - 2k).(j + 3k) = 0 + 0 - 0 - 6 = - 6

(2i - j + 3k).(3i + 2j - k) = 6 - 2 - 3 =1

---

2   IF A = l + 3j - 2k AND B = 4i - 2j + 4k, FIND: (a) A.B, (b) A, (c) B, (d)

|3A + 2B |, (e) (2A + B).(A - 2B).

(a) A.B = (l + 3j - 2k).(4i - 2j + 4k) = 4 - 6 - 8 = - 10

(b) A = (1 + 9 + 4)^(.5) = (14)^.5

(c) B = (16 + 4 + 16)^.5 = 6

(d) | 3A + 2B | = | 3 (l + 3j - 2k) + 2(4i- 2j + 4k) = | 3i + 9j - 6k + 8i - 4j + 8k)

    = |11i - 5j + 2k | = (121 + 25 + 4) ^.5 = (150)^.5

(e) (2A + B).(A - 2B) =

(2A + B) = (2 (l + 3j - 2k) + (4i - 2j + 4k)) = 2i + 6j - 4k + 4i - 2j + 4k

    = (6i + 4j)

(A - 2B) = (l + 3j -2k) - 2(4i - 2j + 4k) = l + 3j -2k - 8i + 4j - 8k

    = -7i + 7J - 10k

(2A + B).(A - 2B) = (6l + 4J).(- 7l + 7J - 10K) = - 42 + 28 = - 14

3. FOR WHAT VALUES OF a ARE A = ai - 2j + k AND B = 2ai + aj - 4k PERPENDICULAR?

A.B = |A||B| cos(θ) = (ai - 2j + k).(2ai + aj - 4k) = 2a^2 - 2a - 4

　　= 2a^2 - 2a - 4 = |A||B| cos (θ) = |A||B| cos (90) = 0

　　　2a^2 - 2^a - 4 = 0

a = (2 + (4 + 32)^.5)/4 = (2 + (6))/4 = 2

a = (2 - (4 + 32)^.5)/4 = (2 - 6)/4 = -1

---

4 FIND THE ACUTE ANGLES WHICH THE LINE JOINING THE POINTS (1, - 3, 2), AND (3, - 5, 1) MAKES WITH THE COORDINATE AXES.

I (y + 3) + k (2)(x - 1) - 2 (z - 2)j + 2(y + 3)k - 2 (z - 2)I - (x - 1)j = 0

I (y + 3 - 2z + 4) + j (4 - 2z - x + 1) + k (2x - 2 + 2y + 6) = 0

Y + 3 -2z + 4 = 0,	4 - 2z - x + 1 = 0,	2x -2 + 2y + 6 = 0
Y = 2z - 7	x = 2z - 5	y = - x - 2
m1 = 2	m2 = 2	m3 = -1

　　　　　m = (2 ^2 + 2^2 + 1) ^.5 = 3

THE ACUTE ANGLES ARE arc cos(2/3),　　arc cos (2/3),　arc cos (1/3)

---

5. ENCUENTRE LOS ANULOS AGUDOS PARA LOS CUALES LA LINEA QUE UNE LOS PUNTOS (1, -3, 2), Y (3, -5, 1) HACE CON LOS EJES COORDENADOS.

I (y + 3) + k (2)(x - 1) - 2 (z - 2)j + 2(y + 3)k - 2 (z - 2)I - (x - 1)j = 0

I (y + 3 - 2z + 4) + j (4 - 2z - x + 1) + k (2x - 2 + 2y + 6) = 0

Y + 3 -2z + 4 = 0,	4 - 2z - x + 1 = 0,	2x - 2 + 2y + 6 = 0
Y = 2z - 7	x = 2z - 5	y = - x - 2
m1 = 2	m2 = 2	m3 = -1

　　　　　m = (2 ^2 + 2^2 + 1) ^.5 = 3

THE ACUTE ANGLES ARE arc cos (2/3),　　arc cos (2/3),　arc cos (1/3)

Y = 2z - 7   x = 2z - 5   y = -x - 2

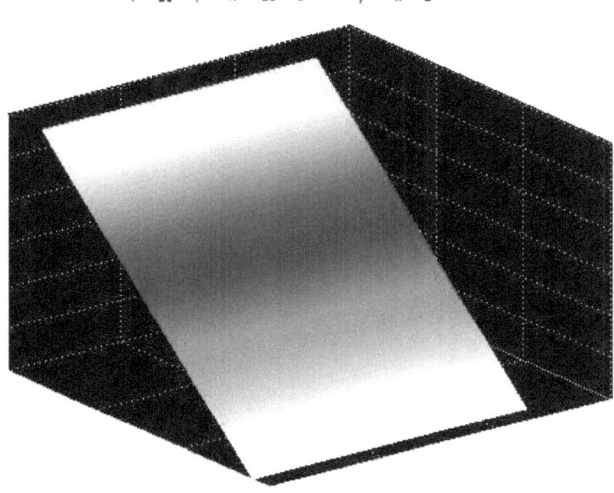

---

6   FIND THE DIRECTION COSINES OF THE LINE JOINING THE POINTS
     (3, 2, -4) AND (1, -1, 2).

r1 = 3i + 2j -4k,   r2 = I - j + 2k, r1.r2 = 3 - 2 -8 = -7

α = (x2 - x1)/r1.r2 = 2/7,   β= (y2 - y1)/r1.r2 = 3/7,   γ= (z1 - z2)/r1.r2 = 6/7

---

7   TWO SIDES OF A TRIANGLE ARE FORMED BY THE VECTORS A = 3i +
     6j - 2k AND B = 4i - j + 3k. DETERMINE THE ANGLES OF THE TRIANGLE.

A.B = 12 - 6 - 6 = 0 THE TRIANGLE IS RECTANGLE.

|A| = (9 + 36 + 4)^.5 = 7, |B| = (16 + 1 + 9)^.5 = (26)^.5

C = (7^2 + 26)^.5 = (49 + 26)^.5 = (75)^.5

Cos α = (26/75)^.5   cos β = 7/(75)^.5   γ = 90 DEGREES

---

8   THE DIAGONALS OF A PARALLELOGRAM ARE GIVEN BY A = 3i - 4j - k
     AND B = 2i + 3j- 6k. SHOW THAT THE PARALLELOGRAM IS A RHOMBUS
     AND DETERMINE THE LENGTH OF ITS SIDES AND ITS ANGLES.

A.B = 6 - 12 + 6 = 0 IS RECTANGLE

|A| = (9 + 16 + 1)^.5 = (26)^.5,

|B | = (4 + 9 + 36) ^.5 = 7

Side = (26 + 49)^.5 = (75)^.5

Cos θ1 = (7)/(75)^.5,   cos θ2 = (26)^.5/(75)^.5 = (26/75)^.5

---

9   FIND THE PROJECTION OF THE VECTOR A = 2i - 3j + 6k ON THE VECTOR
     B = I + 2j+2k.

A.B = 2 - 6 + 12 = 8

|B| = (1 + 4 + 4) ^.5 = 3

A.B = 8 = |A| |B|cos θ = (3) |A| cos θ,

|A| cos θ = 8/3

---

10   FIND THE PROJECTION OF THE VECTOR 4i - 3j + k ON THE LINE
     PASSING THROUGH THE POINTS (2, 3, -1) AND (-2, -4, 3).

A = 4i - 3j + k   B = (x2 - x1)I + (y2 - y1)j + (z2 - z1)k = -4i -7j + 4k;

A.B = -16 + 21 + 4 = 9

|B| = (16 + 49 + 16)^.5 = 9

A.B = |A||B| cos θ = 9 |A| cos θ = 9,

|A| cos θ = 9/9 = 1,

---

11   IF A = 4i - j + 3k, AND B = -2i + j - 2k, FIND A UNIT VECTOR PERPENDICULAR
     TO BOTH A AND B.

$$AxB = \begin{vmatrix} I & j & k \\ 4 & -1 & 3 \\ -2 & 1 & -2 \end{vmatrix} = 2i + 4k - 6j - 2k - 3i + 8j = -I + 2j + 2k$$

AxB = -1 + 2J + 2K, |AxB|= (1 + 4 + 4)^.5 = 3

UNIT VECTOR = (-i+ 2J + 2k)/3

FIND THE ACUTE ANGLE FORMED BY TWO DIAGONALS OF A CUBE.

P1(0,0,1), P2 (1,1,0), P3(0,0,0), P4(1,1,1)

P1P2 = (x2 - x1)i + (y2-y1)j + (z2-z1)k

P!P2 = (1-0)i + (1-0)j + (0-1)k = i + j - k

P3P4 = (1-0)i + (1-0)j + (1-0)k

|P1P2|= (1^2 + 1^2 + 1^2)^.5 = (3)^.5

|P3P4| = (1^1 + 1^2 + 1^2)^.5 = (3)^.5

P1P2. P3P4 = |P1P2||P3P4| cos θ

(1 + 1 - 1) = ((3)^.5) ((3)^.5) cos θ

cos θ = 1/3

   θ = arc cos (1/3)

12 FIND A UNIT VECTOR PARALLEL TO THE xy PLANE AND PERPENDICULAR
    TO THE VECTOR A = 4i - 3j + k.

P1 (x1, y1, 0); VECTOR PARALLEL x1i + x2j + x3k

P1.A = |P1||A|cos θ = 4x1 - 3 y1 = |P1||A| cos (90) = 0

                    Y1/x1 = 4/3 = SLOPE

PERPENDICULAR SLOPE = - 3/4;

Y = mx + b, b = 0 FOR LINE PASSING THROUGH ORIGIN

Y = 3x/4; 4y = 3x OR IN FORM OF UNIT VECTOR r = (3xi - 4yj)/5

13 SHOW THAT A = (2i - 2j + k)/3, B = (I + 2j + 2k)/3 AND C = (2i + j - 2k)/3 ARE
    MUTUALLY ORTHOGONAL UNIT VECTORS.

A.B = (2i - 2j + k). (I + 2j + 2k)/9 = (2 - 4 + 2)/9 = 0 = |A||B| cos θ,
     cos θ = 0, θ = 90 = PI/2,

A.C = (2i - 2j + k).(2i + j - 2k)/9 = (4 -2 - 2)/9 = 0 = |A||B| cos θ,
     cos θ = 0, θ = 90 = PI/2,

C.B = (2i - 2j + k).(2i + j - 2k)/9 = (4 -2 -2)/9 = 0 = |A||B| cos θ,

cos θ = 0, θ = 90 = PI/2,

---

15   FIND THE WORK DONE IN MOVING AN OBJECT ALONG A STRAIGTH LINE FROM (3, 2, -1) TO (2, -1, 4) IN A FORCE FIELD GIVEN BY F = 4i - 3j + 2k.

d = (x2 - x1)I + (y2 - y1)J + (z2 - z1)k = - I - 3J + 5k

T = F.d = (4i - 3j + 2k).(- I - 3J + 5k) = (-4 + 9 + 10) = 15

---

16   LET F BE A CONSTANT VECTOR FORCE FIELD. SHOW THAT THE WORK DONE IN MOVING AN OBJECT AROUND ANY CLOSED POLYGON IN THE FORCE FIELD IS CERO.

F1 = 6i + 0j + 0k   F1 + F2 + F3 = 6i - 3i - J3(3)^.5 - 3i + j3(3)^

F2 = - 3i - j3(3)^.5   F1 + F2 + F3 = 0

F3 = -3i + j3(3)^.5

---

16   PROVE THAT AN ANGLE INSCRIBED IN A SEMI-CIRCLE IS A RIGHT ANGLE.

X1^2 + y1^2 = h1^2

(2R - X1)^2 + y1^2 = h2^2

(2R - X1)^2 = h2^2 - y1^2

2R - x1 = (h2^2 - y1^2)^.5

X1 = 2R - (h2^2 - y1^2)^.5

Sin θ1 = y1/h1 , sin (90 - θ1) = y1/h2

sin (90 - θ1) = cos θ1 = x1/h1

(sin θ1)^2 + (cos θ1)^2 = 1

(Y1/hi) ^2 + (x1/h1)^2 = 1

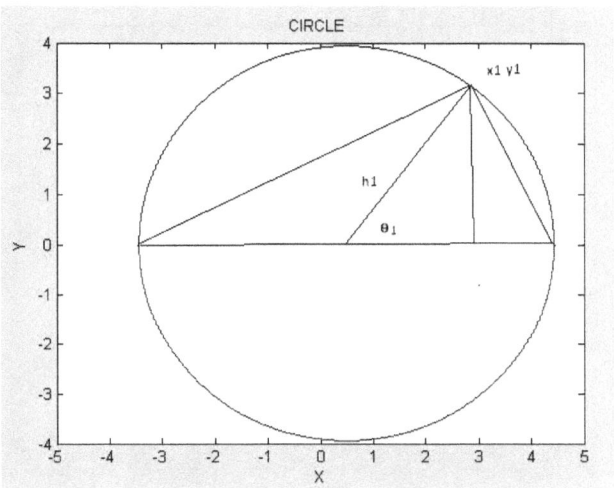

x1^2 + y1^2 = h1^2 IS THE EQUATION OF A SEMI-CIRCLE.

17   LET ABCD BE A PARALLELOGRAM. PROVE THAT AB^2 + BC^2 + CD^2 + DA^2 = AC^2 + BD^2.

AB = iAB Cos θ1 + j AB Sinθ1, BC = Ix1,

CD =- IABCosθ1 - jABSinθ1, DA = - ix1

AB^2 + BC^2 + CD^2 + DA^2 = (ABCosθ1)^2 + (ABSinθ1)^2 + x1^2 + (ABSinθ1)^2

(ABCosθ1)^2 + x1^2 = 2AB^2(Sinθ1^2 + Cosθ1 ^2) + 2 x1^2 = 2(AB^2+ x1^2)

AC = ix1 + iAB cos 1 jAB Sinθ1

BD = ix1+ JABSin 1- iABCosθ1

AC^2 + BD^2 = (ix1 + iABcosθ1 jABSinθ1)^2 + (ix1 + JABSinθ1- iABCosθ1)^2

= x1^2 +(AB cosθ1)^2 + (ABsinθ1)^2+ x1^2 + (ABsinθ1)^2 + θ (ABcosθ1)^2

= 2 x1^2 + 2 AB^2(Sinθ1 ^2 + Cosθ1^2) = 2(AB^2 + x1^2)

PROVED EQUATION   AB^2 + BC^2 + CD^2 + DA^2 = AC^2 + BD^2.

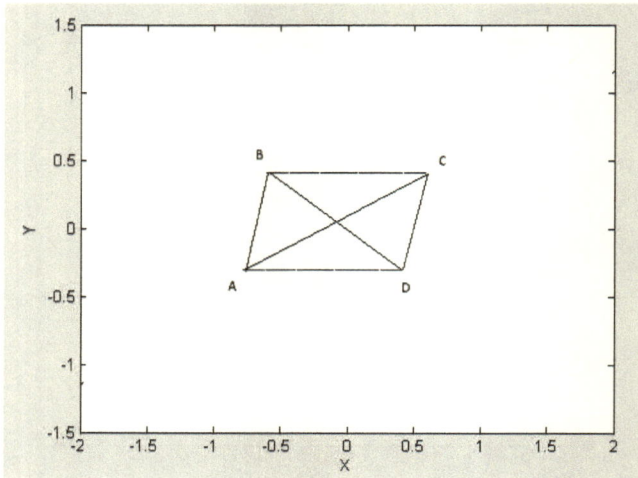

18   (a) FIND AN EQUATION OF A PLANE PERPENDICULAR TO A GIVEN
     VECTOR A AND DISTANT p FROM THE ORIGIN.

UNIT VECTOR $n = A/|A|$;

$r.n = p + |r||n| \cos 90 = p + 0 = p$

$r.n = p$;

(b) EXPRESS THE EQUATION OF (a) IN RECTANGULAR COORDINATES.

$r = xi + yj + zk$; $A = A1i + A2j + A3k$; $n = A/|A|$

$r.n = (xi + yj + zk).(A1i + A2J + A3K)/|A| = p$

$A1x + A2y + A3z = p|A| = pA$

20   LET r1 AND r2 BE UNIT VECTORS IN THE xy PLANE MAKING ANGLES
     AND WITH THE POSITIVE x- AXIS.

(a) PROVE THAT $r1 = \cos \alpha i + \sin \alpha j$, $r2 = \cos \beta i + \sin \beta j$;.

        $r3 = \cos \alpha i + \sin \alpha j$, $r4 = \cos \beta i - \sin \beta j$;

(b) BY CONSIDERING r1.r2 PROVE THE TRIGONOMETRIC FORMULAS

$\cos (\alpha - \beta) = \cos \alpha \cos \beta + \sin \alpha \sin \beta$ $\cos (\alpha + \beta) = \cos \alpha \cos \beta - \sin \alpha \sin \beta$

(a) r1 = cos αi + sin αj GRAPICALLY IS THE DEMOSTRATION

r2 = cosβi + sin βj

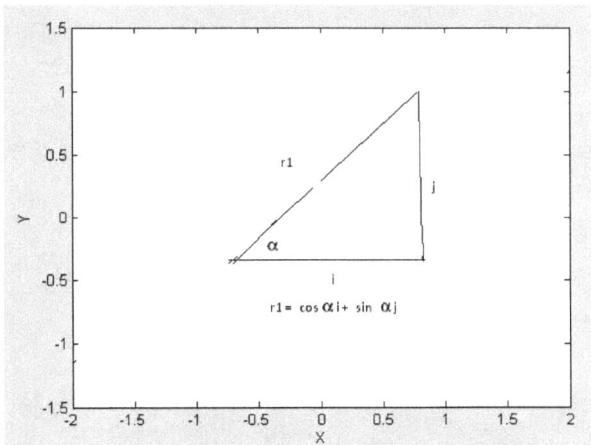

(b)   r1.r2 = cos α cosβ + sin α sin β = |r1||r2| cos (α + β)

   |r1| = ((cos α)^2 + (sin α)^2)^.5 = 1

   |r2| = ((cosβ)^2 + (sin β)^2)^.5 = 1

WE HAVE   cos α cosβ - sin α sin β = cos (α + β)

r3.r4 = cos α cosβ - sin α sin β = |r3||r4| cos (α - β)

   |r3| = ((cos α)^2 + (sin α)^2)^.5 = 1

   |r4| = ((cosβ)^2 + (sin β)^2)^.5 = 1

WE HAVE   cos α cosβ + sin α sin β = cos (α - β)

....

21   LET a b THE POSITION VECTOR OF A GIVEN POINT (x1, x2, x3), AND r

   THE POSITION VECTOR OF ANY POINT (x, y, z). DESCRIBE THE LOCUS

   OF r IF (a) |r - a | = 3,

....

   (b) (r - a).a = 0, (c) (r - a).r = 0.

   (a) r = xi + yj + zk , a = 0i + 0j + 0k,

   |r - a| = |(xi + yj + zk)| = 3

$$= (x^2 + y^2 + z^2)^{.5} = 3$$

$$X^2 + y^2 + z^2 = 9$$

(b) $|r - a|.a = 0$ $a = x1 + y1 + z1;$

$|xi + yj + zk - (x1i + y1j + z1k)|. (x1i + y1j + z1k) = |r - a||a|\cos 90 = 0$

$((x - x1)^2 + (y - y1)^2 + (z - z1)^2)^{.5} = 0/(x1i + y1j + z1k) = 0$

$(x - x1)^2 + (y - y1)^2 + (z - z1)^2 = 0$ IT IS THE EQUATION OF

APERPENDICULAR PLANE TO VECTOR a PASSING THROUGH ITS

TERMINAL POINT.

(c) $(r - a).r = 0.$ $r = xi + yj + zk,$ $a = x1i + y1j + z1k;$

$((xI + yJ + Zk - (x1i + y1j + z1k)).(xi + yj + zk) = 0;$

$((x - x1)i + (y - y1)j + (z - z1) k).(xi + yj + zk) = 0$

$(x - x1) x + (y - y1)y + (z - z1) z = 0$

$(x^2 - x1x + .5 x1^2) + (y^2 - y1y + .5y1^2) + (z^2 - z1z + .5z1^2) = (x1^2 + y1^2$

$+ z1^2) /4$

$(x - X1/2)^2 + (y - y1/2)^2 + (z - z1/2)^2 = (x1^2 + y1^2 + z1^2)/4$

SPHERE WITH CENTER AT (X1/2, Y1/2, Z1/2) AND RADIUS

$r = (x1^2 + y1^2 + z1^2)/2$

22　GIVEN THAT A = 3i + j + 2k, AND B = I - 2j - 4k ARE THE POSITION VECTORS OF POINTS P AND Q RESPECTIVELY. (a) FIND AN EQUATION FOR THE PLANE PASSING THROUGH Q AND PERPENDICULAR TO LINE PQ. (b) WHAT IS THE DISTANCE FROM THE POINT (-1, 1, 1) TO THE PLANE.

$(r - B).(A - B) = 0$

$(xi + yJ + zk - (I - 2j - 4k)).(3i + j + 2k - (I -2j -4k)) =$

$$= | (r - B)||(A - B)|\cos(90) = 0$$

$((x - 1)I + (y + 2)j + (z + 4)k).(2i + 3j + 6k) = 0$

$2(x- 1) + 3(y + 2) + 6(z + 4) = 0$

$2x -2 + 3y + 6 + 6z + 24 = 2x + 3y + 6z + 28 = 0$

EQUATION OF THE PLANE $2 x + 3y + 6z + 28 = 0$

m1 = 2, m2 =3, m3 = 6 m = (2^2 + 3^2 + 6^2)^.5

m = (4 + 9 + 36)^.5 = 7 P (-1, 1, 1)

d = (2x + 3y + 6z +28)/ m = (2(-1) + 3(1) + 6(1) + 28)/7

d = (-2 + 3 + 6 + 28)/7 = 35/7 = 5

d = 5

---

23   EVALUATE EACH OF THE FOLLOWING:

(a) (2j)X(3i - 4k), (b) (I + 2J)X (k), (c) (2i - 4k) X (I + 2J), (d) (4i + j - 2k)X(3i + k), (e) (2i + j - k)X(3i - 2j + 4k)

(a)
$$\begin{vmatrix} I & j & k \\ 0 & 2 & 0 \\ 3 & 0 & -4 \end{vmatrix} = -8I + 0 + 0 - 6k = -8I - 6k$$

(b)
$$\begin{vmatrix} I & j & k \\ 1 & 2 & 0 \\ 0 & 0 & 1 \end{vmatrix} = 2i - j$$

(c)
$$\begin{vmatrix} I & j & k \\ 2 & 0 & -4 \\ 1 & 2 & 0 \end{vmatrix} = 4k - 4j + 8i = 8i - 4j + 4k$$

(d)
$$\begin{vmatrix} i & j & k \\ 4 & 1 & -2 \\ 3 & 0 & 1 \end{vmatrix} = i - 6j - 3k - 4j = i - 10j - 3k$$

(e)
$$\begin{vmatrix} i & j & k \\ 2 & 1 & -1 \\ 3 & -2 & 4 \end{vmatrix} = 4i - 4k - 3j - 3k - 2i - 8j = 2i - 11j - 7k$$

24. IF A = 3i - j - 2k, Y B = 2i + 3j + k, FIND (a) |AxB|, (b) (A + 2B)x (2A - B), (c) | (A + B) x (A - B)|.

(a) $\begin{vmatrix} I & j & k \\ 3 & -1 & -2 \\ 2 & 3 & 1 \end{vmatrix}$ = - I + 9k - 4j + 2k + 6i - 3j = 5I - 7j + 11k

|AxB|= |5i - 7j + 11k| = (25 + 49 + 121)^.5 = (195)^.5

(A + 2B) = (3i -j -2k + 4i + 6j + 2k) = 7i + 5J + 0k

(2A - B) = (6i - 2j -4k - 2i - 3j - k) = 4i - 5j - 5k

(b) $\begin{vmatrix} I & j & k \\ 7 & 5 & 0 \\ 4 & -5 & -5 \end{vmatrix}$ = - 25i - 35k + 0j - 20k + 0i + 35j = -25i - 55k + 35j

(A + B) = (3i - j - 2k + 2i + 3j + k) = 5i + 2j - k

(A - B) = (3i - j - 2k - 2i - 3j -k) = I - 4j - 3k

(c) $\begin{vmatrix} I & j & k \\ 5 & 2 & -1 \\ 1 & -4 & -3 \end{vmatrix}$ = (-6i - 20k - j - 2k - 4i + 15j) = -10i - 22k + 14j

|(A + B) x (A - B)| = ((10)^2 + (22)^2 + (14)^2) ^.5 = (780)^.5 =2(195)^.5

---

25. FIND THE PARALELOGRAM AREA WITH DIAGONALS.

A = 3i + j - 2k, Y B = I - 3j + 4k

$$A \times B = \begin{vmatrix} I & j & k \\ 3 & 1 & -2 \\ 1 & -3 & 4 \end{vmatrix} = 4i - 9k - 2j - (k + 6i + 12j) = -10k - 2i - 14j$$

$|A \times B| = (100 + 4 + 196)^\wedge.5 = (300)^\wedge.5 = 10(3)^\wedge.5/2 = 5(3)^\wedge.5$

26. FIND THE AREA OF A TRIANGLE WITH VERTEXS AT A(3, -1, 2), B(1, -1, -3) y C(4, - 3, 1).

AB = ((1 - 3)^2 + (-1 + 1)^2 + (-3 - 2)^2)^.5 = (4 + 0 + 25)^.5 = (29)^.5

AC = ((4 - 3)^2 + (-3 + 1)^2 + (1 - 2)^2)^.5 = (1 + 4 + 1)^.5 = (6)^.5

BC = ((4 - 1)^2 + (-3 + 1)^2 + (1 + 3)^2)^.5 = (9 + 4 + 16)^.5 = (29)^.5

THE TRIANGLE IS ISOSELES. Hipotenusa = (29)^.5 side1 = (6 /4)^.5 = (3/2)^.5

Side2 = (((29)^.5)^2 - ((3/2)^.5)^2)^.5 = (29 - 3/2)^.5 = ((58 -3)/2) =

(55/2)^.5 Area = (Side1) (side2)/2 = (3/2)^.5 (55/2)^.5 = ((165)^.5)/2

27. SI A = 2i + j - 3k, Y B = I - 2j + k, FIND THE VECTOR OF MAGNITUD 5 PERPENDICULAR BOTH A AND B.

$$A \times B = \begin{vmatrix} I & j & k \\ 2 & 1 & -3 \\ 1 & -2 & 1 \end{vmatrix} = i - 4k - 3j - (k + 6i + 2j) = -5i -5j -5k$$

$|A \times B| = (25 + 25 + 25)^\wedge.5 = (75)^\wedge.5 = 5(3)^\wedge.5$

VECTOR = 5(i + j + k)/(3)^.5 = 5(3)^.5(i + j + k)/3

28   USE PROBLEM 20 TO DERIVE THE FORMULAS

Sin (α+ β) = sin α cosβ - cosαsinβ,

$\sin(\alpha - \beta) = \sin\alpha\cos\beta + \cos\alpha\sin\beta$.

from figure

$r = r\sin\alpha I + r\cos\alpha J$;

$r1 = r1\cos\beta i - r1\sin\beta j$,

$r. r1 = (r\sin\alpha I + r\cos\alpha J)(r1\cos\beta i - r1\sin\beta j) =$

$rr1(\sin\alpha\cos\beta - \cos\alpha\sin\beta) = |r||r1|\cos(\alpha - \beta) = -|r||r1|\sin(\alpha - \beta)$,

$-\sin(\alpha - \beta) = (\sin\alpha\cos\beta - \cos\alpha\sin\beta)$

$\sin(\alpha - \beta) = (-\sin\alpha\cos\beta + \cos\alpha\sin\beta)$

divide both members by $(r\, r1)$

$\cos(\alpha - \beta) = (\cos\alpha\cos\beta - \sin\alpha\sin\beta \,\alpha \,(1)$

$\tan(\alpha - \beta) = (tg\alpha - tg\, s\beta)/(1 + tg\alpha tg\, s\beta) \,(2)$

divide (1) by $\sin(\alpha - \beta)$ we have

$(\cos(\alpha - \beta))(\sin(\alpha - \beta) = (\cos\alpha\cos\beta - \sin\alpha\sin$

$(\alpha - \beta)$ Simplifying

$\cos(-90 + (\alpha - \beta)) = -\sin((\alpha - \beta) = \cos(-90)\cos(\alpha - \beta) - \sin(90)\sin(\alpha - \beta)$

$\sin(\alpha - \beta) = \sin(\alpha - \beta)$

$(\sin(\alpha - \beta)/\cos(\alpha - \beta) = (\sin(\alpha - \beta)/(\cos\alpha\cos\beta - \sin\alpha\sin\beta \,\alpha)$

$\tan(\alpha - \beta) = \sin\alpha\cos\beta + \cos\alpha\sin\beta/(\cos\alpha\cos\beta - \sin\alpha\sin\beta \,\alpha) \,(3)$

From Equetion (2) and (3)

Dividying (3) by $\sin\alpha\cos\beta$- We have

$\tan(\alpha - \beta) = 1 + tg\beta/tg\alpha/(1/tg\,\alpha - \tan\beta) =$

$= (tg\alpha + tg\beta)/(1 - tg\,\alpha\tan\beta)$-

---

28. ONE FORCE GIVEN BY F = 3i + 2j - 4k IS APPLIED AT THE POINT P1(1, -1, 2), FIND THE MOMENT F OV ER THE POINT P2(2, -1, 3):

$F = 3I + 2J - 4K \quad d = (x2 - x1)i + (y2 - y1)j + (z2 - z1)k$

$d = (2 - 1)I + (-1 + 1)J + (3 - 2)k$

$d = I + k$

$$\text{MOMENT} = Fxd = \begin{vmatrix} I & j & k \\ 3 & 2 & -4 \\ 1 & 0 & 1 \end{vmatrix} = 2i + 0 - 4j - 2k + 0 - 3j$$

MOMENT = 2i - 7j -2k

---

29. THE ANGULAR VELOCITY OF A ROTATING RIGID BODY ABOUT AN AXIS OF ROTATION IS GIVEN BY ω = 4I + J - 2K. FIND THE LINEAR VELOCITY OF A POINT P ON THE BODY WHOSE POSITION VECTOR RELATIVE TO A POINT ON THE AXIS OF ROTATION IS 2i - 3j + k.

$$V = \omega \times R = \begin{vmatrix} I & j & k \\ 4 & 1 & -2 \\ 2 & -3 & 1 \end{vmatrix} = I - 12k - 4j - 2k - 6i - 4j = -5i - 8j - 14k$$

---

30. SIMPLIFY (A + B).(B + C)x(C + A) =

(A + B) = X, (B + C) = X, SUPPOSE A = C, C+ A = 2A

X Y X ARE IN THE SAME LINE X.X = |X||X| cosα=

X^2cosαSI α= 0", X^2cos α= X^2 OR IN OTHER WORDS WE HAVE

BXC (A+ B).(B + C) = X^2, Y (C+A) = 2 A (A + B).(B + C)x(C + A) = X^2x(2A) = 2 A.(BxC)

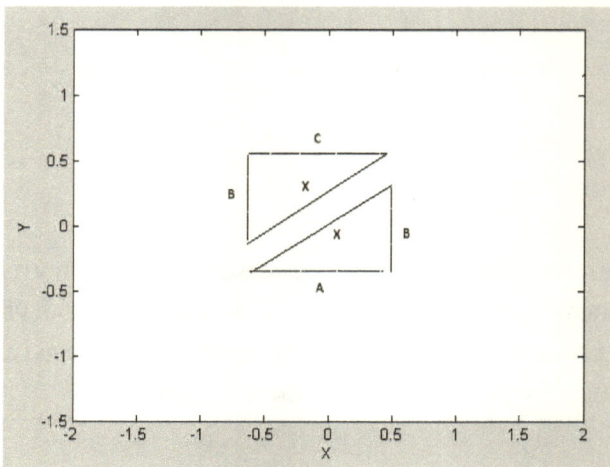

2A.BxC

---

31 PROVE THAT $(A.BxC)(a.bxc) = DET \begin{vmatrix} A.a & A.b & A.c \\ B.á & B.b & B.c \\ C.a & C.b & C.c \end{vmatrix}$

A = Aai + Abj + Ack, B = Bai + Bbj + Bck , C = Cai + Cbj + Cck

$$BxC= \begin{vmatrix} I & j & k \\ Ba & Bb & Bc \\ Ca & Cb & Cc \end{vmatrix} = I\,BbCc + k\,BaCb + CaBcj - (CaBbk + BcCbi + CcBaj)$$

BxC = I(BbCc - BcCb) + j(CaBc - CcBa) + k(BaCb - CaBb)

A.BxC = Aa(BbCc - BcCb) + Ab(CaBc - CcBa) + Ac(BaCb - CaBb)

    = AaBbCc - AaBcCb + AbCaBc - AbCcBa + AcBaCb - AcCaBb

$$Det \begin{vmatrix} Aa & Ab & Ac \\ Ba & Bb & Bc \\ Ca & Cb & Cc \end{vmatrix} = AaBbCc + BaCbAc + CaBcAb - (AcBbCa + BcCbAa + CcBaAb)$$

32. FIND THE VOLUME OF THE PARALLELEPIPED WHOSE EDGES ARE REPRESENTED

BY A = 2I - 3J + 4K. B = i + 2j - k , C = 3i - j + 2k.

V = L. AxH

$$AxH = \begin{vmatrix} I & j & k \\ 1 & 2 & -1 \\ 3 & -1 & 2 \end{vmatrix} = 4i - k - 3j - 6k - I - 2j = 3i - 5j - 7k$$

L.(AxH) = (2i - 3j + 4k).(3i - 5j -7k) = 6 + 15 - 28 = - 7

33. IF A.BxC = 0 SHOWTHAT EITHER (a) A, B AND C ARE COPLANAR BUT TWO OF THEM ARE COLINEAR, OR (b) TWO OF THE VECTORS A, B AND C ARE COLINEAR, OR (c) ALL OF THE VECTORS A, B AND C ARE COLINEAR.

a) A.BxC = 0 IF A.B = |A||B|cosα AND, IF Cos α= cos (90) = 0

b) A.BxC = 0 IF BxC = 0 THE PERPENDICULAR VECTOR IS EQUAL CERO

c) A.BxC = 0 IF A.C = |A||C|Cosα = 0, IF cosα = cos 90" = 0

34. FIND THE CONSTANT a SUCH THAT THE VECTORS A = 2i - j + k, B = I + 2j -3k, Y C = 3i + aj + 5k ARE COPLANR.

A.(BxC) =

$$(BxC) = \begin{vmatrix} I & j & k \\ 1 & 2 & -3 \\ 3 & a & 5 \end{vmatrix} = 10i + ak - 9j - 6k + 3ai - 5j = (10 + 3a)I - (14)j + (a -6)k$$

A.(BxC) = |A||BxC| cos 0; |A| = (2^2 + (-1)^2 + 1^2))^.5 = (4 +1 + 1)^.5

= (6)^.5

|BxC| = ((10 + 3a)^2 + (14)^2 + (a - 6)^2)^.5

A.(BxC) = |A||BxC| cos 0 ;

(2i - j + k).((10 + 3a) I - (14)j + (a -6)k) = ((10 + 3a)^2 + (14)^2 + (a - 6)^2)^.5

(6)^.5)

2(10 + 3a) + 14 + (a - 6) = ((10 + 3a)^2 + (14)^2 + (a - 6)^2)^.5 (6)^.5)

Simplifying both members,

20 + 6a + 14 + a - 6 = 7a + 28 = ((10 + 3a)^2 + (14)^2 + (a - 6)^2)^.5 (6)^.5)

(7a + 28)^2 = 6 (100 + 60a + 9a^2 + 196 + a^2 - 12a + 36) = 49(a + 4)^2

(600 + 360a + 54a^2 + 1176 + 6a^2 - 72a + 216 = 49(a^2 + 8a`+ 16)

60a^2 + 288a - 1392 = 49a^2 + 392a + 784

21a^2 - 104a - 2176 =0

a1 = (104 +(10816 + 182784)^.5)42= (104 - 440)42 = - 8

---

35. LET POINTS P,Q AND R HAVE POSITION VECTORS r1 = 3i - 2j -k , r2 = i +
3j + 4k, r3 = 2i + j - 2k RELATIVE TO AN ORIGEN 0 . FIND THE DISTANCE
FROM P TO THE PLANE OQR.

R2.r3 = 2 + 3 - 8 = - 3 = |(1 + 9 + 16)^.5||(4+ 1 + 4)^.5|cos 1 = (26)^.5(3) cos 1

Cos 1 = -1/(26)^.5 perpendicular slope = (26)^.5

Tg = (m2 - m1)/(1 + m1m2) Tg(90 -) = (m2 - m1)/(1 - m1m2)

    = (26)^.5 - m1)/(1 + m1(26)^.5 = (1 + m1(26)^.5) (-1/(26)^.5 - m1)

    ((26)^.5 - m1) (-1/(26)^.5 - m1) = (1 - 1/(26)^.5 m1) (1 + m1(26)^.5)

    -1 + m1/(26)^.5 - m1(26)^.5 + m1^2 = 1 - 1/(26)^.5m1 - m1(26)^.5 + m1^2

    (26)^.5) + 2m1(1/(26)^.5) -2 = 0

    3.099 = .3922m1 m1 = 7.9

tg x = (5.099 - 7.9)/(1 + (5.099)(7.9)) = -2.801/41.28 =- .067

tg x = - .067, x =- 3" 59 sin - 3" 59 = .0668 = x/h = x/3.74

x = 3.73 y = tg(26)^.5 = 78" 54 z = tg- 1/(26)^.5= - 11" 54

cos (78-4) =x/3.74 , x = .9612(3.74) = 3.6

36   FIND THE SHORTEST DISTANCE FROM (6,-4, 4) TO THE LINE JOINING,
      (2, 1, 2) AND (3, -1,4) .

      center C (6,-4,4)   CIRCLE POINTS A(2,1,2) `AND B(3, -1, 4).   MIDDLE POINT
      AX = XC
      [x - 2 y -1 z - 2] = [3 - x -1 - y 4 - z]
      x- 2 = 3 - x   y - 1 = -1 - y   z -2 = 4 - z
      2x = 5   2y = 0   2z = 6
      X = 5/2   y = 0   z = 3
      R = ((6 - 5/2)^2 + (-4 - 0)^2 + (4 - 3)^2)^.5 = (49/4 + 16 + 1 = (49 + 17(4))/4
      R = (49 + 68)/4 =(1/ 2) (117)^.5 = 5.4
      (x- 6^2 + (Y + 4)^2 + (Z-4)^2 = 117/4

37   FIND A SET OF VECTORS RECIPROCAL TO THE SET 2i + 3j -k, I - j - 2k,
      -i+ 2j + 2k

$$AxB = \begin{vmatrix} I & J & K \\ 2 & 3 & -1 \\ 1 & -1 & -2 \end{vmatrix} = -6i -2k -j -(3k +I -4j) = -9i -5k -5j$$

38   IF a' = bxc/(a.bxc),   b' = cxa/(a.bxc),   c' = axb/(a.bxc) PROVE THAT

      a = b'xc'/(a'.b'xc'),   b = c'xa'/(a'.b'xc'),   c = a'xb'/(a'.b'xc)

      a'= bxc/(a.bxc) = a/(a.a) =1/a, b' = cxa/(a.bxc) = b/(b.axc)= b/a.a
      c' = axb/(a.bxc) = c/a.a
      b'xc' = a',   c'xa' = b',   a'xb' = c'
      a' = (b'. a.a)x(c'a.a)/(a.(b'a.a)x(c'a.a) = a^2.b'xa^2c'/(a.(b'a^2)x(c'a^2)
      a' = b'xc'/a.b'xc', a'.a = (a.b'xc')/(a.b'xc') = b'xc'/b'xc',
      a'.a/a' =b'xc'/a'.b'xc', a = b'xc'/(a'.b'xc')

39   IF a, b, c AND a', b', c' ARE SUCH THAT a'.a = b'.b = c'.c = 1

a'.b = a'.c = b'.a = b'.c = c'.a = c'.b = 0 PROVE THAT IT NECESSARILY

FOLLOW

THAT   a' = bxc/(a.bxc),   b' = cxa/(a.bxc),   c' = axb/(a.bxc)

    Bxc = a   cxa = b   axb = c   a' = a/a.a,   b' = b/a.a,   c' = c/a.a

    a'.a = a.a/a.a = 1,   b'.b = b.b/a.a = 1,   c'.c = c.c/a.a = 1

CHAPTER III

VECTOR AND SCALARS

1   SI R = e^(-t)I + ln (t^2 + 1)j - tan (t) k, ENCUENTRE (a) d R/dt, (b) d R/dt^2, | dR/dt|, (d) | d R/dt^2| AT t = 0

   (a)   dR/dt = - ie^(-t) + j(2t)/(t^2 + 1) – k(sec(t))^2 =- i e^(0) + j (0)/1 – ((cos0)^2/ (sin 0)^2)k = -ie^(0) + k(cos 0)^2/(1 – (cos 0))^2 = - i e^(0) + k(cos 0)^2 / (1 + cos0)(1 –cos 0)

   dR/dt = - i - (1)k

   (b)   dR/dt^2 = i e(-t) + j((2/t^2 +1) – j(2t)^2/(t^2 +1)^2 – k (2sec(t))(sec(t)) (tang(t)

   = ie^(-t) + j(2t^2+ 2) – 2t^2)/(t^2 + 1)^2 – k sec(t)^2 tan(t) = I + j 2

   (c) |dR/dt| = (1+1)^.5 = (2)^.5

   (d) |dR/dt^2 | = (1 + 4)^.5 = 5^.5

2   FIND THE VELOCITY AND ACCELERATION OF A PARTICLE WHICH MOVES ALONG THE THE CURVE x = 2sin 3t, y = 2cos 3t, z = 8t AT ANY TIME t>0, FIND THE MAGNITUDE OF THE VELOCITY AND ACCELERATION.

   x = 2 sin 3t dx/dt = 6cos 3t   dx^2/dt^2 = - 18sin 3t

   Y = 2 cos 3t, dy/dt = - 6sin 3t   dy^2/dt^2 = - 18 cos 3t   dz/dt^2 = 0

   Z = 8t, dz/dt = 8   v = i 6cos 3t – j 6 sin3 t + 8k   a = - 18(isin 3t + jcos 3t)

$|v| = (36 \cos (3t)^2 + 36 (\sin (3t)^2 + 64)^{.5} = (64 + 36(\sin(3t)^2 + \cos 3t)^2) =$

$= (100)^{.5} = 10$

$|a| = 18((\sin 3t)^2 + (\cos 3t)^2) = 18$

---

3   FIND THE UNIT TANGENT TANGENT TO ANY POINT ON THE CURVE.

$x = a \cos wt, \ y = a \sin wt, \quad z = bt \qquad$ WHERE a; b, w ARE CONSTANT.

$dx/dt = - i \ aw \sin wt \qquad dy/dt = j \ a \ w \cos wt, \ dz/dt = kb \ |d/dt| = ((aw)^2 (\sin wt)^2$

$+ \cos wt)^2 + b^2)^{.5} = ((aw)^2 + b^2)^{.5}$

UNIT TANGENT VECTOR $= (- i \ aw \sin wt + j \ aw \cos wt + kb)/((aw)^2 + b^2)^{.5}$

I

dx/dt = - i aw sin wt   dy/dt = j a w cos wt, dz/dt = kb

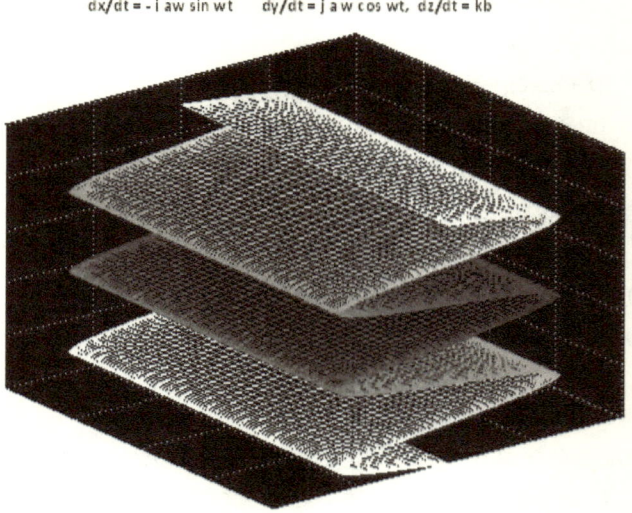

x = a cos wt, y = a sin wt, z = bt

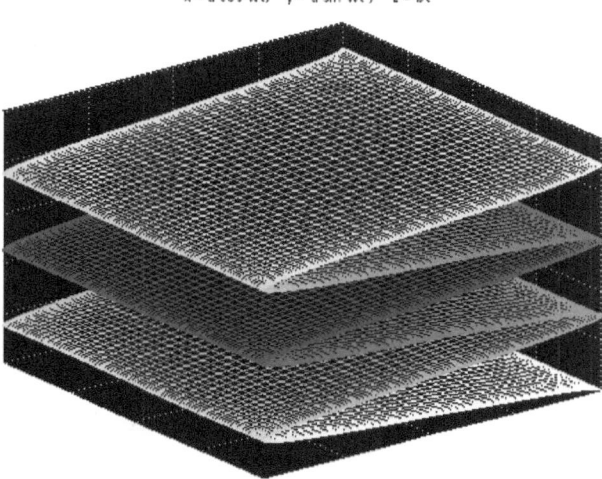

36 FIND d/ds (A.Db/ds - d A/ds. B) IF A AND B ARE DIFFERENTIABLE FUNCTIONS OF s.

d (A.dB/ds - dA/ds.B)/ds = (dA/ds).(dB/ds) + A.d^2B/ds^2 - d^2A/ds^2 -

   - dA/dsdB/ds

   (dA/ds)(dB/ds) = A.(d^2B/ds^2) - (d^2A/ds^2).B

37 FIND (a) THE UNIT TANGENT T, (b) THE CURVATURE κ, (c) THE PRINCIPAL NORMAL N, (d) THE BINORMAL B, AND (e) THE TORSION τ FOR THE SPACE CURVE x = t –

(t^3)/3, y = t^2, z = t + (t^3)/3

$$x = t - (t^3)/3, \quad y = t^2, \quad z = t + (t^3)/3$$

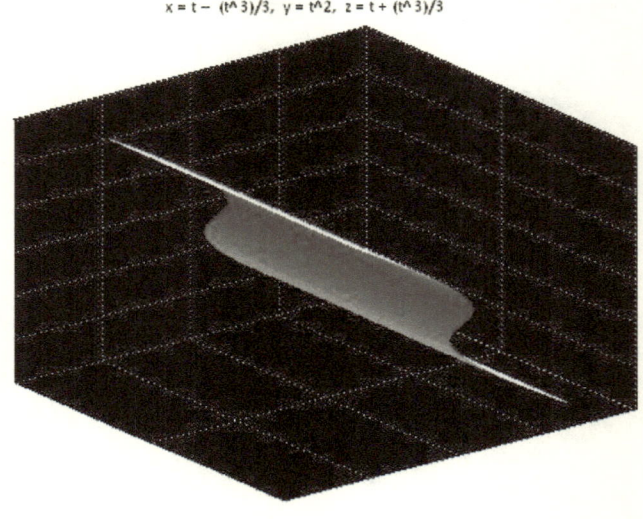

(a)

$dx/dt = 1 - 3(t^2)/3 = 1 - t^2, \quad dy/dt = 2t, \quad dz/dt = 1 + (3t^2)/3 = 1 + t^2$

$dx^2/dt^2 = -2t, \quad dy^2/dt^2 = 2 \quad dz^2/dt^2 = 2t$

$m1 = 1 - t^2, \quad\quad m2 = 2t, \quad\quad m3 = 1 + t^2$

$\quad T = (1 - t^2)I + 2tJ + (1 + t^2)k$

$m = (m1^2 + m2^2 + m3^2)^{.5} = ((1 - t^2)^2 + (2t)^2 + (1 + t^2)^2)^{.5}$

$m = (1 - 2t^2 + t^4 + 4t^2 + 1 + 2t^2 + t^4)^{.5} = (2 + 4t^2 + 2t^4)^{.5} =$

$m = (2(1 + 2t^2 + t^4))^{.5} = (2)^{.5}(1 + t^2)$

$a1 = -2t, a2 = 2, a3 = 2t, a = (4t^2 + 4 + 4t^2)^{.5} = 2(2t^2 + 1)^{.5}$

ax = -2*t; ay = 2; az = 2*t;

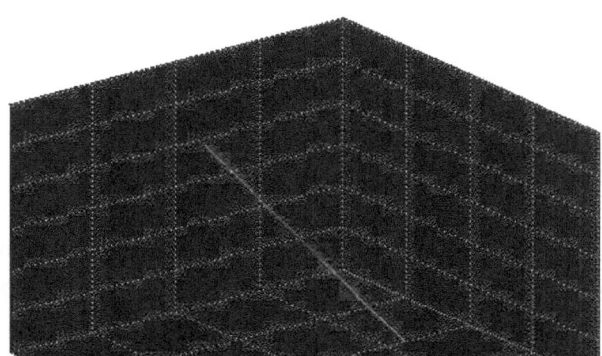

vx = 1 - t.^2; vy = 2*t; vz = 1 + t.^2;

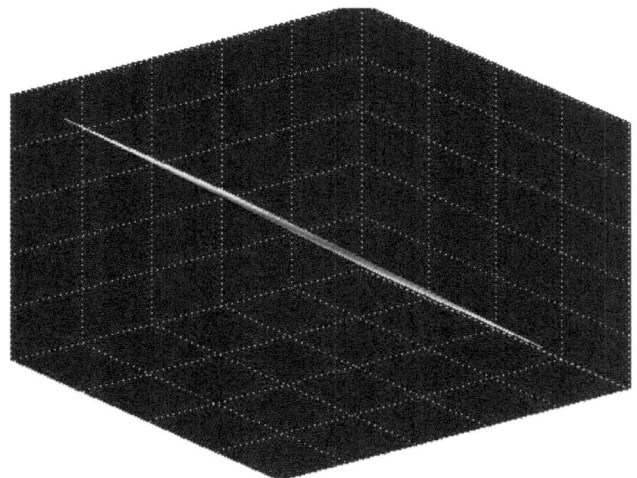

UNIT TANGENT = ((1 − t^2)I + 2tJ + (1 + t^2)k) / (2)^.5(1 + t^2)

(c) Na = 2 (2t^2 + 1) ^.5

T. N = |T||N| cos (90) = 0 = (1 -

n1 = -1/(1 − t^2), n2 = -1/(2t), n3 = -1/(1 + t^2),

N = (- 2ti + 2j + 2tk)

38   d^2A/dt^2 = 6ti - 24t^2j + 4 sin(t) k, FIND A GIVEN THAT A = 2i + j, AND dA/ dt = - I – 3k AT t = 0.

Vx = 6ti, Vy = 24⁺t.^2, Vz = 4⁺ sin(t)

INTEGRATION

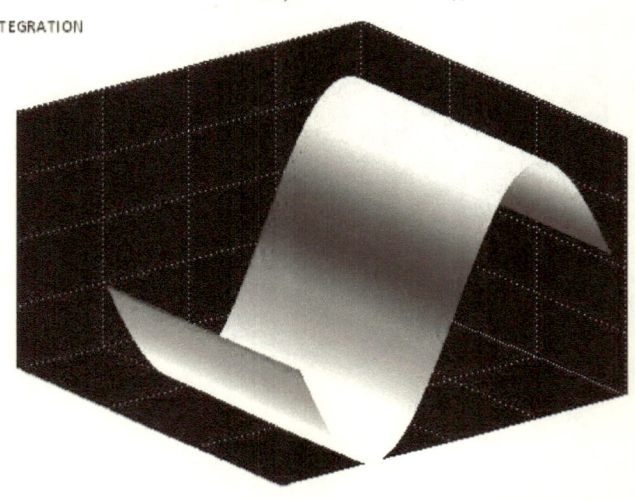

X = (t^3 - t) I - 2 t^4 j + (t - 4 sin t)k

INTEGRATION

$$dA, = 6 \int t \, dti - 24 \int t^2 dtj + 4 \int \sin t \, dtk = 3t^2i - 8t^3j - 4 \cos t \, k + C1$$

d A = - i - 3k = 0i + 0j - 4k + c1

C1 = - I + k

$$A = \int (3t^2 - 1)\, dt_1 \cdot \int 8\, t^3\, dt_j \cdot \int (1 - 4\cos t)\, dt\, k = (t^3 - t)\, I - 2\, t^4 j - (t - 4 \sin t)k - C2 = 2i - j$$

A = (t^3 - t) I - 2 t^4 j + (t - 4 sin t)k + C2 = 2i + j

0 I - 0j + 0k + C2 = 2i + j; C2 = 2i + j

A = (t^3 − t + 2) I - (2 t^4 - 1)j + (t - 4 sin t) k

---

39 SHOW THAT r = e^ (-t) (C1 cos 2t + C2 sin 2t), WHERE C1 AND C2 ARE CONSTANT VECTORS, IS A SOLUTION OF THE DIFFERENTIAL EQUATION d^2 t/dt^2 + 2 dr/dt + 5r = 0.

---

r^2 + 2r + 5 = 0, r = (- 2 +- (4 − 20))/2 = (-2 + - 4i)/2 = -1 +- 2i

r = C1 e^(-1 + 2i) = e^(-t)(C1 cos 2t + C2 sin 2t)

---

40 SHOW THAT THE GENERAL SOLUTION OF THE DIFFERENTIAL EQUATION

---

d r/dt^2 + 2αd r/dt + ω^2 r = 0, where α and ω ARE CONSTANTS, IS

(a) r^2 + 2αd r + ω^2 = 0; r = t(- 2α (4α ^2 - 4ω^2)^.5)/2 = -αt +t (α^2 − ω^2)^.5

(a) r = C1 e^(-αt +t (α^2 - ω^2)^.5) + C2 e^(-αt +t (α^2 - ω^2)^.5)

r = e^(-αt) (C1 e^ (α^2 - ω^2)^.5) t + C2 e^ (- (α^2 - ω^2)^.5) t)) (α^2 - ω^2)^.5) > 0

r = e^(-αt) (C1 sin (α^2 - ω^2)^.5) t + C2 cos (- (α^2 - ω^2)^.5) t)) (α^2 - ω^2)^.5) < 0

r = e^(-αt) (C1 sin(0)t + C2 cos(0)t) = e^(-αt) (C1 + C2 t) (α^2 - ω^2)^.5) = 0

41   SOLVE (a) d r/dt^2 - 4 dr/dt - 5r = 0, (b) dr/dt^2 + 2dr/dt + r = 0, (c) dr/dt^2 + 4r = 0

(a)   r^2 - 4r - 5 = 0 factors (r – 5)(r +1) = 0; roots r1 = 5, r2 = -1

SOLUTION r = C1 e^(5t) + C2 e^(-t)

(b)   r^2 + 2r + 1 = 0 factors (r + 1)^2 = 0 roots r1 = -1, r2 = -1

SOLUTION r = C1e^(-t) + C2 e^(-t) = e^(-t)(C1 + C2t)

(c)   r^2 + 4 = 0 roots r1 = 2e^(-1)^.5, r1 = - 2e^(-1)^.5

SOLUTION r = C1 2e^2((-1)^.5) t + C2 2e^2((-1)^.5)t

FIND d/ds(A.Db/ds - d A/ds. B) IF A AND B ARE DIFFERENTIABLE FUNCTIONS OF s

d   (A.dB/ds - dA/ds.B)/ds = (dA/ds).(dB/ds) + A.d B/ds^2 - d A/ds^2 -

(dA/ds)(dB/ds) = A.(dB/ds^2) - (dA/ds^2).B

37   FIND (a) THE UNIT TANGENT T, (b) THE CURVATURE κ, (c) THE PRINCIPAL NORMAL N, (d) THE BINORMAL B, AND (e) THE TORSION τ FOR THE SPACE CURVE x = t – (t^3)/3, y = t^2, z = t + (t^3)/3

(a)   dx/dt = 1 – 3 (t^2)/3 = 1 - t^2, dy/dt = 2t, dz/dt = 1 + (3t^2)/3 = 1 + t^2

dx/dt^2 = -2t,   dy/dt^2 = 2   dz/dt^2 = 2t

m1 = 1 – t^2,   m2 = 2t,   m3 = 1 + t^2

T = (1 – t^2)I + 2tJ + (1 + t^2)k

m = (m1^2 + m2^2 + m3^2) ^.5 = ((1 – t^2)^2 + (2t)^2 + (1 + t^2)^2)^.5

m = (1 – 2t^2 + t^4 + 4t^2 + 1 + 2t^2 + t^4)^.5 = (2 + 4 t^2 + 2t^4)^.5 =

m = (2(1 + 2t^2 + t^4))^.5 = (2)^.5(1 + t^2)

a1 = -2t, a2 = 2, a3 = 2t, a = (4t^2 + 4 + 4t^2)^.5 = 2(2t^2 + 1)^.5

UNIT TANGENT = ((1 – t^2)I + 2tJ + (1 + t^2)k) / (2)^.5(1 + t^2)

(c)   Na = 2 (2t^2 + 1) ^.5

T. N = |T||N| cos (90) = 0 = (1 -

n1 = -1/(1 – t^2), n2 = -1/(2t), n3 = -1/(1 + t^2),

N = (- 2ti + 2j + 2tk)

39 SHOW THAT r = e^ (-t) (C1 cos 2t + C2 sin 2t), WHERE C1 AND C2 ARE CONSTANT VECTORS, IS A SOLUTION OF THE DIFFERENTIAL EQUATION d t/dt^2 + 2 dr/dt + 5r = 0.

r^2 + 2r + 5 = 0,   r = (- 2 +- (4 − 20))/2 = (-2 + - 4i)/2 = -1 +- 2i

r = C1 e^(-1 + 2i) = e^(-t)(C1 cos 2t + C2 sin 2t)

40 SHOW THAT THE GENERAL SOLUTION OF THE DIFFERENTIAL EQUATION

d r/dt^2 + 2αd r/dt + ω^2 r = 0, where α and ω CONSTANTS, IS

r^2 + 2αd r + ω^2 = 0; r = (- 2α (4α^2 - 4ω^2)^.5)/2 = -α + (α^2 - ω^2)^.5

r = C1 e^(-α + (α^2 - ω^2)^.5) + C2 e^(-α + (α^2 - ω^2)^.5)

r = e^(-α) (C1 e^ (α^2 - ω^2)^.5) t + C2 e^ (- (α^2 - ω^2)^.5) t))(α^2 - ω^2)^.5)

   > 0

r = e^(-α) (C1 sin (α^2 - ω^2)^.5) t + C2 cos (- (α^2 - ω^2)^.5) t))

(α^2 − ω^2)^.5) < 0

r = e^(-α) (C1 sin(0)t + C2 cos(0)t = e^(-α) (C1 + C2 t)

(α^2 - ω^2)^.5) = 0

41 SOLVE (a) d r/dt^2 - 4 dr/dt - 5r = 0, (b) dr/dt^2 + 2dr/dt + r = 0, (c) dr/dt^2 + 4r = 0

(a) r^2 - 4r - 5 = 0 factors (r − 5)(r +1) = 0; roots r1 = 5, r2 = -1
     SOLUTION r = C1 e^(5t) + C2 e^(-t)
(b) r^2 + 2r + 1 = 0 factors (r + 1)^2 = 0 roots r1 = -1, r2 = -1
     SOLUTION r = C1e^(-t) + C2 e^(-t) = e^(-t)(C1 + C2t)
(b) r^2 + 4 = 0 roots r1 = 2e^(-1)^.5, r1 = - 2e^(-1)^.5
     SOLUTION r = C1 2e^2((-1)^.5) t + C2 2e^2((-1)^.5)t
     SOLUTION r = C1 sin 2t + C2cos 2t

42   SOLVE dY/dt = X, dX/dt = - Y.

r – 1 = 0	r + 1 = 0
r = 1	r = -1
X = e^(t) = C1 cos t + C2 sin t	- Y = e^(-t) = C1 cos (-t) + C2 sin (-t)
	- Y = C1cos(t) - C2 sin(t)
	Y = C2 sin (t) - C1 cos (t)

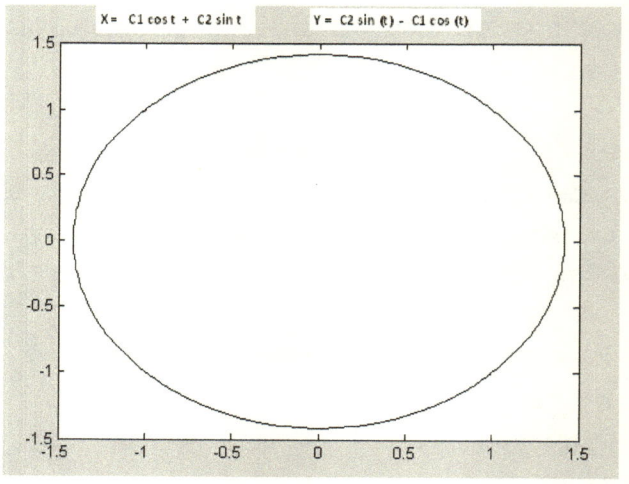

dY/dt = X   dX/dt = -Y

dt = dY/X = - dX/Y

YdY = - XdX   Integrating we have Y^2/2 + X^2/2 = C or

X^2 + Y^2 = C   circle equation.

---

43   IF A = icos xy + (3xy – 2x^2)j - (3x + 2y)k.

FIND (a) δA/ δx, δA/ δy, δA/ δx^2, δA/ δy^2, δA/ δx δy,

δA/ δyδx. -----------

δA/ δx = - i ysinxy + (3y – 4x)j - (3)k

δA/ δy = -i xsin xy + 3xj - 2k

δA/ δx^2 = -i (y^2)cos xy – 4j

$\delta A/ \delta y^2 = -i (x^2) \cos xy$

$\delta A/ \delta x \delta y = - i \, yx \cos xy - \sin xy + 3j$

$\delta A/ \delta y \delta x = - I \, xy \cos xy - \sin xy + 3j$

---

44   IF $A = (x^2)yz \, I - 2 \, xz^3 \, j + xz^2 \, k$ AND $B = 2z \, I + yj - (x^2)k$, FIND $\delta(AxB)/$
$\delta x \delta y$ AT (1, 0, -2). ---------

$AxB = \begin{vmatrix} I & J & k \\ (x^2 yz & -2xz^3 & xz^2 \\ 2z & Y & -X^2 \end{vmatrix} = 2i(xz)^3 + z(xy)^2k + 2\,xz^3j + 4xz^4k - xyz^2i + yz(x^4)j$

$AxB = I \, (2(XZ)^3 - XY \, Z^2) + J \, (2XZ^3 + YZ \, X^4) + k \, (2Z \, (XY)^2 + 4 \, XZ^4)$

$\delta(AxB)/\delta x = I \, (6 \, (X^2)(Z^3) + YZ^2) + j \, (2Z^3 + 4 \, YZ \, X^3) + k \, (4 \, XZY^2 + 4Z^4)$

$\delta(AxB)\delta x \delta y = I \, (Z^2) + j \, (4 \, ZX^3) + k \, (8 \, XYZ)$ AT POINT (1, 0, -2)

$\delta(AxB)\delta x \delta y = - 4 \, I - 8j$

---

45   IF C1 AND C2 ARE CONSTANT VECTORS AND IS A CONSTANT SCALAR,
SHOW THAT $H = e(-\lambda x)(C1 \sin \lambda y + C2 \cos \lambda x)$ SATISFIES THE PARTIAL
DIFFERENTIAL EQUATION

$\delta H/\delta x^2 + \delta H/\delta y^2 = 0$

$\delta H/ \delta y = \lambda e(-\lambda x)(C1 \cos \lambda y)$

$\delta H/ \delta y^2 = - (\lambda^2) e(-\lambda x) (C1 \sin \lambda y)$

$\delta H/\delta x = -\lambda e(-\lambda x)(C1 \sin \lambda y + C2 \cos \lambda x) + \lambda e(-\lambda x)(- C2 \sin \lambda x)$

$\delta H/\delta x = - \lambda e(-\lambda x)(C1 \sin \lambda y + C2 \cos \lambda x) - \lambda e(-\lambda x)(C2 \sin \lambda x)$

$= - \lambda \, e(-\lambda x) \, (C1 \sin \lambda y + C2 \cos \lambda x - C2 \sin \lambda x) =$

$\delta H/\delta x = (\lambda) \, e(-\lambda x)(C1 \sin \lambda y) - C2 \, \lambda \, e(-\lambda x) \, (\cos \lambda x - \sin \lambda x) = (\lambda) \, e(-\lambda x)(C1$
$\sin \lambda y)$

FOR EQUAL ANGLES $C2 \, \lambda \, e(-\lambda x) \, (\cos \lambda x - \sin \lambda x) = 0$

$\delta H/\delta x^2 = (\lambda^2) \, e(-\lambda x)(C1 \sin \lambda y)$

WE DWMOSTRATE THAT $\delta H/\delta x^2 + \delta H/\delta y^2 = 0$

CHAPTER IV

GRADIENT

---

44 IF F = (x^2)z + e^(y/x) AND G = 2 (z^2)y – x(y^2), FIND (a) ∇(F + G), AND ∇ (FG) AT THE POINT (1,0,-2).

(F + G) = (x^2)z + e^(y/x) + 2 (z^2)y – x(y^2),

(a) ∇ (F + G) = (ẟ(F + G)/ ẟx)i + (ẟ(F + G)/ ẟy)j + (ẟ(F + G)/ ẟz)k

= (2xz + e^(y/x) (- y/x^2) + y^2)i + (e^(y/x) (1/x) + 2z^2 + 2xy)j +

(x^2 + 4zy)k   in the point (1,0,-2).

∇ (F + G) = (- 4)i + (1 + 8)j + (1)k = -4i + 9j + k

(b) ∇ (FG) = ((x^2)z + e^(y/x))(2 (z^2)y – x(y^2)) =

∇ (FG) = (ẟ(FG)/ ẟx)i + (ẟ(FG)/ ẟy)j + (ẟ(FG)/ ẟz)k

(FG) = 2(x^2)(z^3)y + 2y(z^2)(e^(y/x)) - (x^3)(y^2)z –

(e^(y/x))xy^2

(ẟ(FG)/ ẟx)i = 4 x(z^3)y + (2y(z^2)(e^(y/x)(- y/x^2) – (3x^2)(y^2)z – (Y^2(e^(y/x) +

Xy^2((e^(y/x))(-y/X^2) in the point (1,0,-2)

(ẟ(FG)/ ẟx)i = 0i

(ẟ(FG)/ ẟy)j = 2(x^2)(z^3) + 2(z^2(e^(y/x) + 2y(z^2)(e^(y/x))(1/x) – (x^3)z(2y)

- (e^(y/x))(2xy) – xy^2(e^y/x)(1/x) at the point (1,0.-2)

(ẟ(FG)/ ẟy)j = 2(1)^2(-2)^3 + 2(4)(1) + 0 – 0 -0 -0 = (-16 + 8)j = -8j

(ẟ(FG)/ ẟz)k = 0

∇ (FG) = -8j

45   FIND ∇ |r^3|

$r = xi + yj + zk$, | r | = $(x^2 + y^2 + z^2)^{.5}$ |r^3| = $(x^2 + y^2 + z^2)^{(3/2)}$

∇|r^3| = (δ|r^3|/ δx)i + (δ|r^3|/ δy)j + (δ|r^3|/ δz)k

$((3/2)(x^2 + y^2 + z^2)^{(1/2)})(2X)$

δ|r^3|/ δx)i = $3x (x^2 + y^2 + z^2)^{(1/2)}$ i

(δ|r^3|/ δy)i = $3y (x^2 + y^2 + z^2)^{(1/2)}$ j

(δ|r^3|/δz)i = $3z (x^2 + y^2 + z^2)^{(1/2)}$ k

∇|r^3| = $3x (x^2 + y^2 + z^2)^{(1/2)}$ I + $3y (x^2 + y^2 + z^2)^{(1/2)}$ j

+ $3z (x^2 + y^2 + z^2)^{(1/2)}$ k

47   EVALUATE ∇ (3 r^2 - 4 (r)^.5 + 6 (r) ^(-1/3))

$r = xi + yj + zk$, r.r = $(x^2 + y^2 + z^2)$, 3 r^2 = $3(x^2 + y^2 + z^2)$

$|r| = (x^2 + y^2 + z^2)^{.5}$, 4|r| = $4 (x^2 + y^2 + z^2)^{.5}$

$6/(|r^{(1/3)} = 6 / ((x^2 + y^2 + z^2)^{(1/6)}$

(3 r^2 - 4 (r)^.5 + 6 (r) ^(-1/3)) = $3(x^2 + y^2 + z^2) - 4 (x^2 + y^2 + z^2)^{.5} + 6 / ((x^2 + y^2 + z^2)^{(1/6)}$

(3 r^2 - 4 (r)^.5 + 6 (r) ^(-1/3)) = $3(x^2 + y^2 + z^2) - 4 (x^2 + y^2 + z^2)^{.5} + 6 / ((x^2 + y^2 + z^2)^{(1/6)}$

∇ (3 r^2 - 4 (r)^.5 + 6 (r) ^(-1/3)) = i δ /δX + jδ /δY + k δ/δz

i δ /δX = $6x - 4 (x^2 + y^2 + z^2)^{(-3/2)} - (2x/(x^2 + y^2 + z^2)^{(7/2)})$

jδ /δY = $6y - 4y (x^2 + y^2 + z^2)^{(-3/2)} - (2y/(x^2 + y^2 + z^2)^{(7/2)})$

k δ/δz = $6z - 4z (x^2 + y^2 + z^2)^{(-3/2)} - (2z/(x^2 + y^2 + z^2)^{(7/2)})$

∇ (3 r^2 - 4 (r)^.5 + 6 (r) ^(-1/3)) = $(6 - 4 r^{(-3/2)} + 2r^{(-7/2)})$ r

48   IF ∇ U = 2 r^4 R = 2(|r|^4)r, FIND U.

∇ (U) = i δu/δX + jδu /δY + k δu/δz

r = xi + y j + zk, | r | = (x^2 + y^2 + z^2)^.5, |r|^4 = (x^2 + y^2 + z^2)^2

∇ (U) = i δu/δX + jδu /δY + k δu/δz = 2 (x^2 + y^2 + z^2)^2(xi + y j + zk)

i δu/δX = 2 xi (x^2 + y^2 + z^2)^2,

iδu = 2 i (x^2 + y^2 + z^2)^2(x) Δx, w = (x^2 + y^2 + z^2)

Dw = 2xdx,

i|dU = i w^2 dw =   iUx = I w^3/3

Ux = i((x^2 + y^2 + z^2)^3)/3 THE SAME WAY

WE HAVE

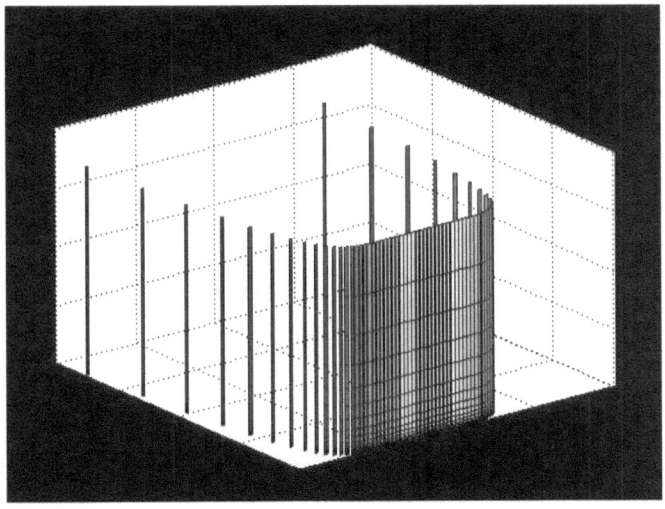

Uy = ((x^2 + y^2 + z^2)^3) /3

Uz = ((x^2 + y^2 + z^2)^3)/3

iUx + jUy + k Uz = (|r|^6)/3 + CONSTANTE

49   FIND (r) SUCH THAT ∇ (U) = r/|r|^5 AND U (1) = 0.

r = xi + yj + zk, | r| = (x^2 + y^2 + z^2)^.5, | r|^5 = (x^2 + y^2 + z^2)^(5/2)

∇ (U) = iδu/δX + jδu /δY + k δu/δz

i δUx = (xi + yj + zk)/(x^2 + y^2 + z^2)^(5/2)/ i δx

δux = i xδx /(x^2 + y^2 + z^2)^(5/2), w = (x^2 + y^2 + z^2)

dw = 2x dx

uX = i dw/(2(w)^(5/2))

ux = - I (2/3) /(2(w)^(3/2)) = -I/(3(x^2 + y^2 + z^2)^(3/2)) + c1

ux(1) = -1/3 +C1 = 0, C1 = 1/3

uy = - j/(3(x^2 + y^2 + z^2)^(3/2)) + c2

uy(1) = -1/3 + C2 =0, C2 = 1/3

uz = - k /(3((x^2 + y^2 + z^2)^(3/2)) +C3

uz(1) = -1/3 + C3=0, C3 = 1/3

U = ux + uy + uz = 1/3(1 – 1/r^3)

$$U = 1/3(1 - 1/r^3)$$

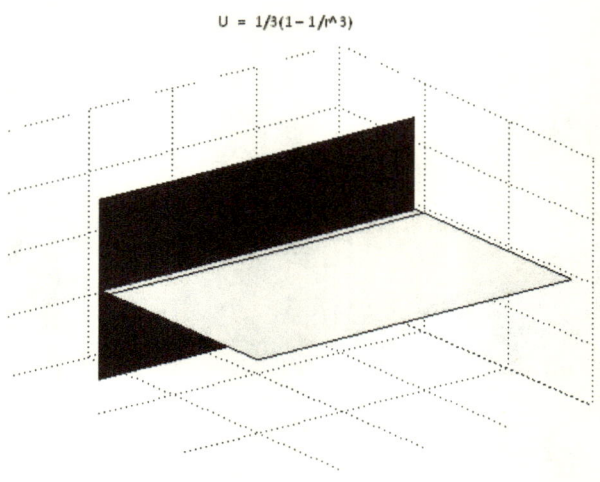

---

50   FIND ∇ψ WHERE ψ = (x^2 + y^2 + z^2) e^(- (x^2 + y^2 + z^2)^.5)

∇ (ψ) = i δψ/δX + jδψ/δY + k δψ/δz

i δψ/δX = 2x e^(- (x^2 + y^2 + z^2)^.5) - x(x^2 + y^2 + z^2)^(1/2) e^

(- (x^2 + y^2 + z^2)^.5)

$= ixe^{\wedge}(- (x^2 + y^2 + z^2)^{\wedge}.5) (2 - | r|)$

$j \, \delta\psi/\delta y = 2y \, e^{\wedge}(- (x^2 + y^2 + z^2)^{\wedge}.5) - y(x^2 + y^2 + z^2)^{\wedge}(1/2) \, e^{\wedge}($

$- (x^2 + y^2 + z^2)^{\wedge}.5)$

$= jye^{\wedge}(- (x^2 + y^2 + z^2)^{\wedge}.5) (2 - | r|)$

$k \, \delta\psi/\delta z = 2z \, e^{\wedge}(- (x^2 + y^2 + z^2)^{\wedge}.5) - z(x^2 + y^2 + z^2)^{\wedge}(1/2) \, e^{\wedge}(- ($

$x^2 + y^2 + z^2)^{\wedge}.5)$

$= kze^{\wedge}(- (x^2 + y^2 + z^2)^{\wedge}.5) (2 - | r|)$

$\nabla (\psi) = ((2 - |r|)e^{\wedge}|r|)r$

---

51   IF $\nabla \varphi = 2xyz^3 \, i + x^2z^3 \, j + 3x^2yz^2 \, k$. FIND $\varphi(x, y, z)$ IF $(1, -2, 2) = 4$

$\nabla (\varphi) = i \, \delta\varphi/\delta X + j\delta\varphi/\delta Y + k \, \delta\varphi/\delta z$

$\delta\varphi x/\delta X = 2xyz^3$,	$\delta\varphi x = 2xyz^3\delta x$,	$\varphi x = x^2yz^3 + C1$
$\delta\varphi y/\delta Y = x^2z^3$,	$\delta\varphi y = x^2z^3 \, \delta Y$,	$\varphi y = yx^2z^3 + C2$
$\delta\varphi z/\delta z = 3x^2yz^2$,	$\delta\varphi z = 3x^2yz^2\delta z$	$\varphi z = yx^2z^3 + C3$

$\varphi = \varphi x + \varphi y + \varphi z = yx^2z^3 + C4$ AT THE POINT $(1, -2, 2)$

$(-2) (1)^2 (2)^3 + C4 = 4$

$-16 + C4 = 4$

$C4 = 20$

$\varphi = yx^2z^3 + 20$

---

52   IF $\nabla (\psi) = (x^2 - 2XYZ^3)I + (3 + 2XY - X^2Z^3)j + (6z^3 - 3x^2yz^2)k$, FIND ψ.

$\nabla (\psi) = i \, \delta\psi/\delta X + j\delta\psi/\delta Y + k \, \delta\psi/\delta z$

$\delta\psi x/\delta X = 2xyz^3 + x^2$, $\delta\psi x = (2xyz^3 + x^2) \, \delta x$, $\psi x = x^2yz^3 + (x^3)/3$

$\delta\psi y/\delta Y = 3 + 2XY - X^2Z^3$, $\delta\psi y = (3 + 2XY - X^2Z^3)\delta Y$, $\psi y = 3y + xy^2 - yx^2z^3 + C2$

$\delta\psi z/\delta z = (6z^3 - 3x^2yz^2)$, $\delta\psi z = (6z^3 - 3x^2yz^2)\delta z$

$\psi z = (3z^4)/2 - x^2yz^3 + C3$ $\psi = \psi x + \psi y + \psi z = x^2yz^3 + (x^3)/3 + 3y + xy^2 - yx^2z^3 + (3z^4)/2 + x^2yz^3 = - x^2yz^3 + xy^2 + 3y + (3z^4)/2 + C$

$\psi = - x^2yz^3 + xy^2 + 3y + (3z^4)/2 + \text{constant}$

53   IF U IS A DIFFERENTIABLE FUNCTION OF x, y, z, PROVE ∇ U.dr = dU.

∇ U = i δu/δX + jδu/δY + k δu/δz

r = xi + yj + zk, dr = dxi + dyj + dzk

∇ U.dr = (i δu/δX + jδu/δY + k δu/δz).(dxi + dyj + dzk) =

∇ U.dr = (δu/δX) dx + (δu/δY) dy + (δu/δz) dz = δu

---

54   IF F IS A DIFFERENTIABLE FUNCTION OF x, y, z, t WHERE x, y, z ARE DIFERENTIABLE

FUNCTION OF t, PROVE THAT dF/dt = δF/δt + ∇ F. dr/dt.

∇ F. dr/dt r = xi + yj + zk, dr/dt = idx/dt + j dy/dt + k dz/dt

∇ F = i δF/δX + jδF/δY + k δF/δz

∇ F. dr/dt = (δF/δX) dx/dt + (δF/δY) dy/dt + (δF/δz) dz/dt

∇ F. dr/dt = dF/dt + dF/dt + dF/dt + δF/δt = dF/dt

---

55   IF A IS A CONSTANT VECTOR, PROVE ∇ (r.A) = A

r = xi + yj + zk, A = Axl + AyJ + Azk

∇ (r.A) = i δ(r.A)/δX + jδ(r.A)/δY + k δ(r.A)/δz = Ax + Ay + Az = A

---

56   IF A(x, y, z) = A1i + A2j + A3k, SHOW THAT dA = (∇A1.dr)I + (∇ A2.dr)j +

(∇ A3.dr)k.

r = xi + yj + zk, dr = dxi + dyj + dzk, I∇A1.dr = I(dA1dx)/dx = IdA1

∇(A2.dr) J = J(dA2dx)/dx, = Jd A2, K∇(A3.dr) = K(dA3dz)/dz = kdA3

dA = dA1i + dA2j + dA3k

---

57   PROVE ∇(F/G) = (G∇F - F∇G)/(G^2)

∇ (F/G) = (∇F)/G - F(∇G)/ G^2 = (G∇F - F∇G)/(G^2)

58 FIND A UNIT VECTOR WHICH IS PERPENDICULAR TO THE SURFACE
OF THE PARABOLOID OF REVOLUTION z = x^2 + y^2 AT THE POINT (1,
2, 5).

φ = x^2 + y^2 - z

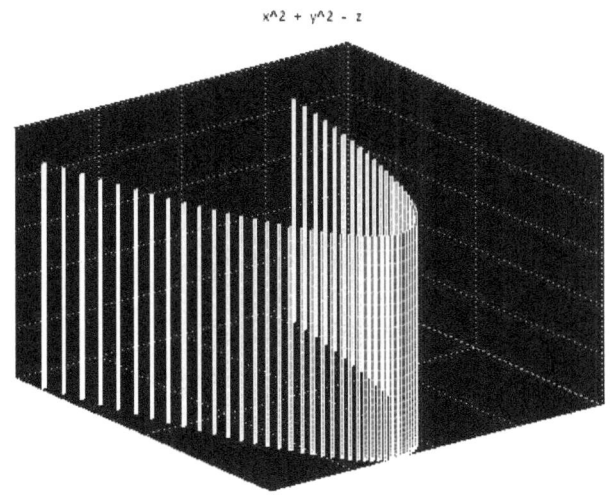

δφ/δx = 2x, δφ/δy= 2y δφ/δz = -1 at the point (1, 2, 5)

δφ/δx = 2   δφ/δy= 4   δφ/δz = -1 Perpendicular Slope

δφ/δx = m1 = -1/ 2, δφ/δy = m2 =- 1/4 δφ/δz = m3 = 1 straigth line
equations

y -2 = (-1/ 2)(x – 1) y – 2 = (-1/4) (x – 1) y – 2 = 1(z – 5)

$y - 2 = (-1/2)(x - 1)$ $y - 2 = (-1/4)(x - 1)$ $y - 2 = 1(z - 5)$

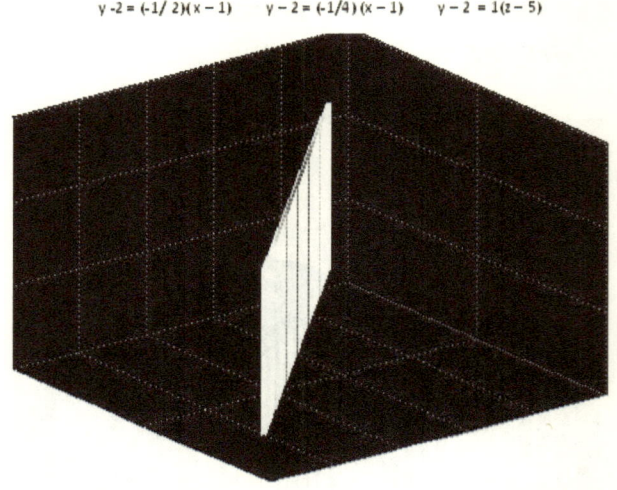

$\delta\varphi/\delta x = 2$ $\delta\varphi/\delta y = 4$ $\delta\varphi/\delta z = -1$

unit vector = $(2i + 4j - k)/(4 + 16 + 1)^{.5} = (2i + 4j - k)/(21)^{.5}$

---

59   FIND THE UNIT OUTWARD DRAWN NORMAL TO THE SURFACE $(x - 1)^2$ $+ y^2 + (z + 2)^2 = 9$ AT THE POINT (3, 1, -4).

$\varphi = (x - 1)^2 + y^2 + (z + 2)^2 - 9$

$\delta\varphi/\delta x = 2(x - 1) = 2(3 - 1) = 4$, $\delta\varphi/\delta y = 2y = 2$,

$\delta\varphi/\delta z = 2(z + 2) = 2(-4 + 2) = -4$

unit outward = $(4i + 2j - 4k)/(16 + 4 + 16)^{.5} = (4i + 2j - 4k)/6 = (2i + j - 2k)/3$

---

60   FIND AN EQUATION FOR THE TANGENT PLANE TO THE SURFACE xz^2 $+ (x^2)y = z - 1$ AT THE POINT (1, -3, 2).

= $xz^2 + (x^2)y - z + 1$

$\delta\varphi/\delta x = z^2 + 2xy = (2)^2 + 2(1)(-3) = 4 - 6 = -2$,

$\delta\varphi/\delta y = (x^2) = 1$

$\delta\varphi/\delta z = 2xz - 1 = 2(1)(2) - 1 = 4 - 1 = 3$

- 2x + y + 3z + C = 0 at the point (1, -3, 2) we have

- 2(1) - 3 + (3)(2) + C = 0,

$$C = -1$$

2x – y - 3z + 1 = 0

61 FIND EQUATION FOR THE TANGENT PLANE AND NORMAL LINE TO THE SURFACE

Z = x^2 + y^2 AT THE POINT (2, -1, 5)

φ = x^2 + y^2 - z

δφ/δx = 2x = 4, δφ/δy = 2y = -2, δφ/δz = -1

(x - 2)/4 = t,	(y + 1)/(-2) = t	(z - 5)/(-1) = t
x = 4t + 2	y = -2t -1	z = - t + 5

4x - 2y - z + cte = 0,

4(2) – 2(-1) - 5 + CTE = 0

Cte = - 8 – 2 + 5 = -5

4x - 2y - z = 5

62 FIND THE DIRECTIONAL DERIVATIVE OF P = 4xz^3 – 3z(xy)^2 AT THE POINT (2, -1, 2) IN THE DIRECTION 2i - 3j + 6k.

δφ/δx = 4z^3 - 6 zxy^2 = 4(2)^3 – 6(2)(2)(1) = 32 – 24 = 8

δφ/δy = - 6 zyx^2 = - (6)(2)(-1)(2)^2 = 48

δφ/δz = 12xz^2 - 3(xy)^2 = 12(2)(2)^2 - 3(2)^2 = 96 - 12 = 84

(2i - 3j + 6k)/(4 + 9 + 36)^.5 = (2i - 3j + 6k)/ 7

Directional derivative = (8i + 48j + 84k)(2i – 3j + 6k)/7 = 16/7 - 144/7 + 504/7

Directional derivative = 376/7

63   FIND THE DIRECTIONAL DERIVATIVE OF P =4 e^(2x - y + z) AT THE
POINT (1, 1, -1) IN A DIRECTION TOWARD THE POINT (-3, 5, 6).

$\delta P/\delta x$ = 8e^ (2x –y +z) = 8e^(2 -1 - 1) = 8e^(0) = 8

$\delta P/\delta y$ = - 4e^(2x –y + z) = - 4e^(0) = -4

$\delta \varphi/\delta z$ = 4e^(2x – y + z) = 4e^(0) = 4 AT THE POINT -3i + 5j + 6k

Directional derivative = (-3i + 5j + 6k)(8i -4j + 4k)/((70)^.5(96)^5) =-24 – 20

+24 = -20/9

CHAPTER V

VECTOR INTEGRATION

28 IF $R(t) = (3t^2 - t)i + (2 - 6t)j - 4tk$, FIND (a) $\displaystyle\int_{0}^{} R(t)\,dt$ AND $\displaystyle\int_{2}^{4} R(t)\,dt$

$(t^3 - (t^2)/2)i + (2t - 3t^2)j - (2t^2)k$

$[(t^3 - (t^2)/2)i + (2t - 3t^2)j - (2t^2)k] = (64 - 8)i + (8 - 48)j - 32k$

$- (8 - 2)i - (4 - 12)j + 8k = (56)i - 40j - 32k - 6i + 8j + 8k = 50i - 32j - 24k$

29 EVALUATE $\displaystyle\int_{0}^{\pi/2} (3\sin u\, i + 2\cos u\, j)\,du$

$= [-3\cos u]i + [2\sin u]j = 3i + 2j$

30 IF $A(t) = ti - t^2 j + (t - 1)k$ $B(t) = (2t^2)i + 6tk$ EVALUATE (a)

$$\int A.Bdt \quad (b) \quad \int AxBdt$$

A.B = (2t^3) + 6(t-1)t = (2t^3) + 6(t^2 – t)

$$AxB = \begin{vmatrix} I & j & k \\ 2t^2 & 0 & 6t \\ t & -t^2 & (t-1) \end{vmatrix} = -2kt^4 + 6jt^2 + 6it^3 - 2jt^2(t-1)$$

AxB = (6t^3)I + (- 2t^3 +8t^2)j - (2t^4)k

$$\int A.Bdt$$

$$= \int_0^2 (2 t^3 + 6t^2 - 6t\ )dt = [(t^4)/2 + (2t^3 - 3t^2)] = (8 + 16 - 12) = 12$$

= [t^2(t^2/2 +2t -3)] = 4(3) = 12

(b) $$\int_0^2 AxB = \int_0^2 (\ (6t^3)I\ +\ (- 2t^3 +8t^2)j\ -\ (2t^4)k\ )\ dt$$

= [(3/2)t^4i –(1/2)j t^4 + (8/3)jt^3 – (2/5)kt^5] = 24i – (8 - 64/3)j – (k64/5)

= 24i + (40/3)j – k(64/5)

32 THE ACELERATION a OF A PARTICLE AT ANY TIME t >= 0 IS GIVEN BY a = e^ (-t) I – 6(t + 1)j + 3(sin t)k. IF THE VELOCITY v AND DISPLACEMENT r ARE ZERO AT t = 0, FIND v AND r AT ANY TIME.

$$v = \int a\,dt = \int (e^{(-t)}I - 6(t + 1)j + 3(\sin t)k)dt$$

= - e^(-t)I – (3t^2 + 6t)j - (3 cos t)k)dt + cte

= - i - 3k + cte = 0, cte = i + 3k

V = (1 − e^(-t)) I - (3t^2 + 6t)j + (3 - 3cost)k

$$r = \int v\,dt = \int ((1 - e^{(-t)})I - (3t^2 + 6t)j + (3 - 3cost)k)dt$$

r(t) = (t + e^(-t) I - (t^3 + 3t^2)j + (3t - 3sinu)k + cte

r(0) = i + cte = 0, cte = - i

r(t) = (-1 + t + e^(-t))i - (t^3 + 3t^2)j + (3t - 3sinu)k

33 THE ACCELERATION a OF AN OBJECT AT ANY TIME t IS GIVEN BY a = - gt WHERE g IS A CONSTANT. AT t = 0, THE VELOCITY IS GIVEN BY v = v0 cos θ0 I + v0 sin θ0j. AND THE DISPLACEMENT r = 0. FIND v AND r AT ANY TIME t > 0. THIS DESCRIBE THE MOVEMENT OF PROJECTILE FIRED FROM A CANNON INCLINED AT ANGLE θ0 WITH POSITIVE x AXIS WITH INITIAL VELOCITY OF MAGNITUDE v0.

$$v = \int - g\,dt$$

v(t) = -gtj + cte, v(0) = -gtj + cte = cte

V(t) = (v0) cos θ0 I + (-gt + v0) sin θ0 j

$$r(t) = \int v\,dt$$

= (v0)t cos θ0i + ((-gt^2)/2 + v0t) sin θ0j + cte

r(0) = 0 + cte, cte = 0

r(t) = (v0)t cos θ0i + ((-gt^2)/2 + v0t) sin θ0j

---

34   EVALUATE $\displaystyle\int_{2}^{3}$ AdAdt/dt   IF A(2) = 2i − j + 2k, AND A(3) = 4i - 2j + 3k

---

$$\int_{2}^{3} A\cdot dA$$

= [(A^2)/2] = (A(3)^2 - A(2)^2)/2 = (16 + 4 + 9 - (4 + 1 + 4))/2

= 20/2 = 10

---

35   FIND THE AREAL VELOCITY OF A PARTICULE WHICH MOVES ALONG THE PATH

---

r = a cosωti + b sinωtj WHERE a, b AND ω ARE CONSTANTS AND t IS TIME. (acosωti)(b sinωtj) = ab (sinωt) (cosωt)k = kab (sin2ωt)/2

Sin2ωt = 2 sinωt cosωt, ωt = u, ωdt = du, k ab sin u (ωdu)/2

$$\int k\ ab\ \sin u\ (\omega du)/2$$

= kabω (- cos u) /2] = kabω (- cos90 + cos 0) /2

= k abω/2

36 PROVE THAT THE SQUARES OF THE PERIODS OF PLANETS IN THEIR MOTION AROUND THE SUN ARE PROPORTIONAL TO THE CUBES OF THE MAJOR AXES OF THEIR ELLIPTICAL PATHS (KEPLER THIRD LAW).

$(x/a)^2 + (y/b)^2 = 1$, $(bx)^2 + (ay)^2 = (ab)^2$

$(br\cos\theta)^2 + (ar\sin\theta)^2 = (ab)^2$, $r^2((b\cos\theta)^2 + (ar\sin\theta)^2) = (ab)^2$

$r = (ab)/ ((b\cos\theta)^2 + (ar\sin\theta)^2))^{.5}$

Period $= (ab)d\theta/((b\cos\theta)^2 + (ar\sin\theta)^2))^{.5}$

37 IF A = (2y + 3)I + xy j + (yz − x) k, EVALUATE Adr ALONG THE FOLLOWING PATHS C:

(a) $x = 2t^2$, $y = t$, $z = t^3$ FROM t = 0, TO t = 1,

(b) THE STRAIGTH LINES FROM (0, 0, 0) TO (0, 0, 1), THEN TO (0, 1, 1) AND THEN TO (2, 1, 1).

(c) THE STRAIGTH LINE JOINING (0, 0, 0) AND (2, 1, 1).

dr = dx + dy + dz, dx = i(2t + 3)dt, $x = [(t^2 + 3t)I] = [1 + 3]I = 4i$

41 EVALUATE F.dr WHERE F = (x − 3y)I + (y − 2x) j AND C IS THE CLOSED CURVE IN THE xy PLANE, x = 2cos t, y = 3 sin t FROM t= 0 TO t = 2π.

dr = dxI + dyJ + dzk, F = (x − 3y)i + (y − 2x)j, dx = 2cost dt

F.dr = (x − 3y) dx + (y − 2x)dy $(\sin t)^2 = (1 + \cos (2t))/2$

F.dr = - 2(2cos t − 9 sin t)sin t dt + 3(3 sin t − 2 cos t)cost dt

$= - 4 (\cos t)(\sin t)dt + 18 (\sin t)^2 dt + 6(\cos t)^2 dt + 9 \sin t \cos t\ dt$

$= - 4((\cos t)^2 + \sin t)^2)dt - 9\sin (2t) dt + 13 (\sin t)^2 dt$

$- 3\sin (2t)\ dt$

$= - 4dt − 12 \sin (2t) + 13 (\sin t)^2 dt$

$= -4dt - 12 \sin(2t)\ dt + (5/2) (1 + \cos (2t))\ dt$

$= - 4t + 6\cos(2t) + (5/2)t + (5/4)\sin(2t)$

$= [- 8\pi + 6 \cos (4\pi) + 5\pi + (5/4)\sin (4\pi) - 0 - 6 \cos(0) -0 -$

(5/4) sin (0)]

$= -3\pi + 6 + 0 - 6 - 0 = -3\pi$

42　IF T IS A UNIT TANGENT VECTOR TO THE CURVE C, r = r(u), SHOW THAT THE WORK DONE IN MOVING A PARTICLE IN A FORCE FIELD F ALONG C IS GIVEN BY F. T ds WHERE s IS THE ARC LENGTH.

43　IF F = (2x + y^2)I + (3y − 4x) j, EVALUATE F.dr AROUND THE TRIANGLE C OF FIGURE 1, (a) IN THE INDICATED DIRECTION, (b) OPPOSITE DIRECTION.

r1 = 2I, r2 = 2i + J, r3 = 0I + 0J, dr = dxi + dyj + dzk

direction 1 F.dr = (2x + y^2)dx + (3y − 4x)dy = [x^2 + xy^2 + (3y^2)/2

- 4xy] = 4

direction 2 F.dr = [x^2 + xy^2 + (3y^2)/2 - 4xy] = 4 + 2 + 3/2 − 8 =

$= -2 + 3/2 = -\frac{1}{2}$

direction 3 F. dr =

44　IF F = (2x^2 − 3z)i − 2xyJ - 4Xk, EVALUATE (a) $\iiint \nabla_x F dV,$

$\iiint \nabla.F dV$ AND (　(b) WHERE V IS THE CLOSED REGION

BOUNDED BY THE PLANES x = 0, y = 0, z = 0 AND 2x + 2y + z = 4.

$\nabla.F = \delta/\delta x (2x^2 -3z) - \delta/\delta y (2xy) - \delta/\delta z (4x) = 4x − 2x =$

$$\int\int\int\limits_{x=0}^{z=4-2x-2y} 2x\ dx\ dy\ dz$$

$$\int\int\limits_{y=0}^{z=0}$$

$$[2xz]_{0}^{4-2x-2y} = (2x(4-2x-2y)) = \int\int\limits_{x=0}^{y=2-x}(8x-4x^2-4xy)\,dx\,dy$$

$$[8xy-4x^2y-2xy^2]_{0}^{2-x/2} = 8x(2-x) -4x^2(2-x) -2x(2-x)^2$$

$$= \int_{0}^{2}(8x-8x^2-2x^3)\,dx$$

$$= [4x^2-8x^3/3-x^4/2] = 16-64/3-8 = 8/3$$

$$\nabla.F = \begin{vmatrix} i & j & k \\ \delta/\delta x & \delta/\delta y & \delta/\delta z \\ 2x^2-3z & -2xy & -4x \end{vmatrix}$$

Iδ/δy(- 4x) + k δ/δx(- 2xy) + j δ/δz(2x^2 -3z) - k δ/δy(2x^2 -3z) - j

δ/δx(- 4x) - iδ/δz(- 2xy) = i(0) + j(-3 + 4) + k(-2y) = j – 2yk +0i

(b)

$$= j \int_{x=0} \int_{y=0} \int_{z=0}^{z=4-2x-2y} dx\,dy\,dz$$

$$[z]_{0}^{4-2x-2y} = 4-2x-2y \quad = j \int_{x=0} \int_{y=0}^{y=2-x} (4-2x-2y)\,dx\,dy$$

$$= [4y - 2xy - y^2]_{0}^{2-x} = 4(2-x) - 2x(2-x) - (2-x)^2 = 8 - 4x - 4x + 2x^2 - (4 - 4x + x^2) =$$

$$= j \int_{x=0}^{x=2} (4 - 4x + x^2)\,dx$$

$$= j \int_{x=0}^{x=2} (4 - 4x + x^2)\,dx = J[4x - 2x^2 + x^3/3]\ 0\ 2 = j(8 - 8 + 8/3) = j(8/3)$$

$$= -k \int_{x=0} \int_{y=0} \int_{z=0}^{z=4-2x-2y} 2y\,dx\,dy\,dz$$

$$-[2yz]_0^{4-2x-2y} = (2y(4-2x-2y)) = -k \int_{y=0} \int_{x=0}^{x=2-y} (8y-4y^{\wedge}2 - 4xy)\,dy\,dx =$$

$$= -[(8yx - 4y^{\wedge}2x - 2x^{\wedge}2y)]_0^{(2-y)} = 8y(2-y) - 4y^{\wedge}2(2-y) - 2y(2-y)^{\wedge}2 =$$

$$= -(16y - 8y^{\wedge}2 - 8y^{\wedge}2 + 4y^{\wedge}3 - 8y + 8y^{\wedge}2 - 2y^{\wedge}3) = -(8y - 8y^{\wedge}2 + 2y^{\wedge}3)$$

$$= -k \int^{y=2} (8y - 8y^{\wedge}2 + 2y^{\wedge}3)\,dy$$

$$= -k(4y^{\wedge}2 - 8y^{\wedge}3/3 + y^{\wedge}4/2) = -k(16 - 64/3 + 8) = -k\,8/3$$

CHAPTER VI

37   DETERMINE THE TRANSFORMATION FROM (a) SPHERICAL TO RECTANGULAR COORDINATES, (b) SPHERICAL TO CYLINDRICAL.

(a)   SPHERICAL TO RECTANGULAR COORDINATES

$X = r \sin\theta \cos\varphi \qquad y = r\sin\theta\sin\varphi \qquad z = r\cos\theta$

$x^2 + y^2 + z^2 = (r \sin\theta \cos\varphi)^2 + (r\sin\theta\sin\varphi)^2 + (r\cos\theta)^2$

$= (r \sin\theta)^2 ((\cos\varphi)^2 + (\sin\varphi)^2) + (r\cos\theta)^2$

$X^2 + y^2 + z^2 = (r \sin\theta)^2 + (r\cos\theta)^2 = r^2((\sin\theta)^2 + (\cos\theta)^2)$

$X^2 + y^2 + z^2 = r^2$

(b)   SPHERICAL TO CYLINDRICAL.

$X = r \sin\theta \cos\varphi \qquad y = r\sin\theta\sin\varphi \qquad z = r\cos\theta$

TRANSFORMATION $\rho = r \sin\theta$, $z = r\cos\theta$

$X = \rho\cos\varphi$, $y = \rho\sin\varphi$, $z = z$

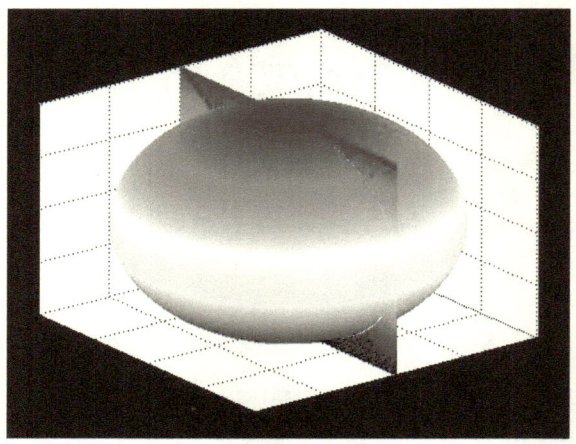

38 EXPRESS EACH OF THE FOLLOWING LOCI IN SPHERICAL COORDINATES:

a) THE SPHERE $x^2 + y^2 + z^2$

$R^2 (\sin\theta)^2$

$X = r \sin\theta \cos\varphi \qquad y = r\sin\theta\sin\varphi \qquad z = r\cos\theta$

$(r \sin\theta \cos\varphi)^2 + (r\sin\theta\sin\varphi)^2 + (r\cos\theta)^2 = 9$

$(r \sin\theta)^2 ((\cos\varphi)^2 + (\sin\varphi)^2) + (r\cos\theta)^2 = 9$

$(r^2)((\sin\theta)^2 + (\cos\theta)^2) = 9$

$r = 3$

$x = 3 \sin\theta \cos\varphi \qquad y = 3\sin\theta\sin\varphi \qquad z = 3\cos\theta$

b) THE CONE $z^2 = 3 (x^2 + y^2)$

$(3\cos\theta)^2 = (3 \sin\theta \cos\varphi)^2 + (3\sin\theta\sin\varphi)^2$

$((3\cos\theta)^2)/ (3 \sin\theta)^2 = (\cos\varphi)^2 + (\sin\varphi)^2 = 1$

$1/(\tan \theta)^2 = 1, \tan \theta = 1, \theta = 45 = \pi/4$

c) THE PARABOLOID $z = x^2 + y^2$

$X = r \sin\theta \cos\varphi \qquad y = r\sin\theta\sin\varphi \qquad z = r\cos\theta$

$r\cos\theta = (r \sin\theta \cos\varphi)^2 + (r\sin\theta\sin\varphi)^2$

$r\cos\theta = (r \sin\theta)^2 ((\cos\varphi)^2 + (\sin\varphi)^2) = (r \sin\theta)^2$

$r\cos\theta = (r\sin\theta)^2 = r^2 (\sin\theta)^2$

$\cos\theta = r(\sin\theta)^2$

d)   THE PLANE z = 0,

$z = r\cos\theta = 0$

$\cos\theta = 0$, $\theta = 90 = \pi/2$

e)   THE PLANE y = x

$X = r \sin\theta \cos\varphi$ $y = r\sin\theta\sin\varphi$

$r\sin\theta\sin\varphi = r \sin\theta \cos\varphi$

$\tan \varphi = 1$, $\varphi = 45 = \pi /4$

39   IF ρ, φ,z ARE CYLINDRICAL COORDINATES, DESCRIBE EACH OF THE FOLLOWING LOCI AND WRITE THE EQUATION OF EACH LOCUS IN RECTANGULAR COORDINATES:

(a) ρ = 4, z = 0;

$X = r \sin\theta \cos\varphi$ $y = r\sin\theta\sin\varphi$ $z = r\cos\theta$

TRANSFORMATION $\rho = r \sin\theta$, $z = r\cos\theta$

$X = \rho\cos\varphi$, $y = \rho\sin\varphi$, $z = z$

$X = 4 \cos\varphi$, $y = 4 \sin\varphi$, $z = z$

$X^2 + y^2 = (4 \cos\varphi)^2 + (4 \sin\varphi)^2 = 16((\cos\varphi)^2 + (\sin\varphi)^2) = 16$

$X^2 + y^2 = 16$, $z = 0$ CIRCLE IN THE xy PLANE

(b) ρ= 4

$X^2 + y^2 = (4 \cos\varphi)^2 + (4 \sin\varphi)^2 = 16((\cos\varphi)^2 + (\sin\varphi)^2) = 16$

$X^2 + y^2 = 16$, CYLINDER WHOSE AXIS COINCIDE WITH THE z AXIS.

(c) φ = π/2

$X = \rho\cos\varphi$, $y = \rho\sin\varphi$, $x = \rho\cos(\pi /2)$, $y = \rho\sin(\pi /2)$,

$X = 0$, $y = \rho$ THE yx PLANE WHERE y >= 0.

(d) φ = π/3, z = 1.

$X = \rho\cos\varphi$, $y = \rho\sin\varphi$, $x = \rho\cos(\pi /3)$, $y = \rho\sin(\pi /3)$, $z = 1$

$X = \rho\cos(\pi /3) = \rho\cos(60) = \rho(1/2)$, $y = \rho\sin(\pi /3) = \rho\sin(60)$

$Y = \rho((3)^{.5})/2 = ((3)^{.5})x$, $z = 1$, THE STRAIGTH LINE $y = ((3)^{.5})x$, $z =1$

41   IF u,v,z ARE PARABOLIC CYLINDRICAL COORDINATES, GRAPH THE
      CURVES OR REGIONS DESCRIBED BY EACH OF THE FOLLOWING:

a)   U = 2, z = 0;   zk = (xi + yj + zk)cos

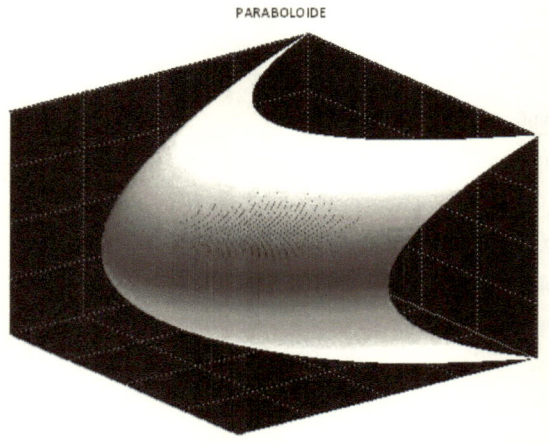

PARABOLOIDE

$X = (u^2 - v^2)/2$, y = uv, z = z

$X = (4 - v^2)/2$, y = 2v, z = 0

$X = (4 - (y^2)/4)/2$, z = 0

$2x = (16 - y^2)/4$, $8x = (16 - Y^2)$, $Y^2 = 16 - 8X = 8(2 - X)$

Z = 0,

b)   u = 1, z = 2

$x = (1 - v^2)/2$, y = v,

$2x = (1 - y^2)$, z = 2

42   FIND THE UNIT VECTORS er, eθ, AND eφ OF A SPHERICAL COORDINATE
      SYSTEM IN TERMS OF I, j AND k. (b) SOLVE FOR I, j AND k IN TERMS OF
      er, eθ, AND eφ.

$X = r \sin\theta \cos\varphi$ y = $r\sin\theta\sin\varphi$ z = $r\cos\theta$

r = xi + yj + zk = ir sinθ cosφ + jrsinθsinφ + k rcosθ

δr/δr = i sinθ cosφ + jsinθsinφ + k cosθ

|δr/δr| = (X^2 + y^2 + z^2)^.5 =

=((sinθ cosφ)^2 + (sinθsinφ)^2 + (cosθ)^2)^.5

= ((sinθ)^2((cosφ)^2 + (sinφ)^2)) + (cosθ)^2))

= ((sinθ)^2 + (cosθ)^2) = 1

er = δr/δr/ |δr/δr| = i sinθ cosφ + jsinθsinφ + k cosθ

δr/δφ = -i sinθ sinφ + jsinθcosφ

|δr/δφ|= ((sinθ sinφ)^2 + (sinθcosφ)^2)^.5 =

= ((sinθ)^2((sinφ)^2 + cosφ)^2))^.5 = sinθ

eφ = δ/δφ/|δ/δφ|= - i sinφ + j cosφ

r = xi + yj + zk = ir sinθ cosφ + jrsinθsinφ + k rcosθ

δr/δθ = ir cosθ cosφ + jrcosθsinφ - kr sinθ

|δr/δθ| = ((rcosθ cosφ)^2 + (rcosθsinφ)^2 - (r sinθ)^2)^.5

= ((rcosθ)^2 ((cosφ)^2 + (sinφ)^2)) + (r sinθ)^2)^.5

= ((rcosθ)^2 + (r sinθ)^2)^.5 = r

eθ= δr/δθ/|δr/δθ| = i cosθ cosφ + jcosθsinφ - k sinθ

REPRESENT THE VECTOR A = 2yi - zj + 3xk IN SPHERICAL COORDINATES
AND DETERMINE Ar, Aθ AND Aφ.

X = r sinθcosφ, y = r sinθsinφ, z = r cosθ

A = 2irsinθsinφ - rj cosθ + 3rk sinθcosφ

eφ = - i sinφ + jcosφ eθ = i cosθ cosφ + jcosθsinφ –

    k sinθ

er = isinθ cosφ + jsinθsinφ + k cosθ

er = isinθ cosφ + jsinθsinφ + k cosθ

eθ = i cosθ cosφ + jcosθsinφ - k sinθ

er + eθ = i cosφ (sinθ + cosθ) + j sinφ(sinθ + cosθ) + k(cosθ

    - sinθ)

k = er + eθ/(cosθ - sinθ) - j sinφ(sinθ + cosθ) / (cosθ - sinθ)

– i cosφ (sinθ + cosθ)/ (cosθ - sinθ)

PROVE THAT SPHERICAL COORDINATE SYSTEM IS ORTOGONAL.

$X = r \sin\theta\cos\varphi$, $y = r \sin\theta\sin\varphi$, $z = r \cos\theta$

$r = xi + yj + zk = r i\sin\theta\cos\varphi + rj\sin\theta\sin\varphi + r k\cos\theta$

$\delta r/\delta r = i\sin\theta\cos\varphi + j\sin\theta\sin\varphi + k\cos\theta$

$\delta r/\delta\theta = ir \cos\theta \cos\varphi + jr\cos\theta\sin\varphi - kr \sin\theta$

$\delta r/\delta\varphi = - i \sin\theta \sin\varphi + j\sin\theta\cos\varphi$

$er = \delta r/\delta r/ |\delta r/\delta r| = i \sin\theta \cos\varphi + j\sin\theta\sin\varphi + k \cos\theta$

$e\theta = \delta r/\delta\theta/|\delta r/\delta\theta| = i \cos\theta \cos\varphi + j\cos\theta\sin\varphi - k \sin\theta$

$e\varphi = \delta r/\delta\varphi/|\delta r/\delta\varphi| = - i \sin\varphi + j \cos\varphi$

$er.e\theta = \sin\theta \cos\theta (\cos\varphi)^2 + \sin\theta \cos\theta (\sin\varphi)^2 - \sin\theta \cos\theta$

$er.e\theta = \sin\theta \cos\theta ((\cos\varphi)^2 + (\sin\varphi)^2) - \sin\theta \cos\theta = 0$

$er.e\varphi = (i \sin\theta \cos\varphi + j\sin\theta\sin\varphi + k \cos\theta). (- i \sin\varphi + j \cos\varphi)$

$er.e\varphi = - \sin\theta \cos\varphi \sin\varphi + \sin\theta\sin\varphi \cos\varphi = 0$

$e\theta .e\varphi = (i \cos\theta \cos\varphi + j\cos\theta\sin\varphi - k \sin\theta). (- i \sin\varphi + j \cos\varphi)$

$e\theta .e\varphi = -\cos\theta \cos\varphi \sin\varphi + \cos\theta\sin\varphi \cos\varphi = 0$

THE SYSTEM IS ORTHOGONAL.

47 EXPRESS THE VELOCITY v AND ACCELERATION a OF A`PARTICLE IN SPHERICAL COORDINATES.

$X = r \sin\theta\cos\varphi$, $y = r \sin\theta\sin\varphi$, $z = r \cos\theta$

$r = xi + yj + zk = r i\sin\theta\cos\varphi + r j\sin\theta\sin\varphi + rk \cos\theta$

$vr = \delta r/\delta t$, $v\theta = r \delta\theta/\delta t$, $v\varphi = r \sin\theta \delta\varphi/\delta t$,

$v = er\, vr + e\theta\, v\theta + e\varphi v\varphi$

$ar = \delta^2 r/\delta t^2$, $a\theta = r \delta^2\theta/\delta t^2$, $a\varphi = r \sin\theta \delta\varphi^2/\delta t^2$,

$ar = \delta^2 r/\delta t^2$, $a\theta = \delta^2 r\delta t^2 \delta\theta/\delta t + r \delta\theta/\delta t^2 +$

$a\varphi = \sin\theta \delta\varphi/\delta t \delta r/\delta t + r (\sin\theta \delta\varphi/\delta t^2 + \cos\theta \delta\theta/\delta t$
      $\delta\varphi/\delta t)$

$a\varphi = \sin\theta \delta\varphi/\delta t \delta r/\delta t + r\sin\theta \delta\varphi/\delta t^2 + r\cos\theta \delta\theta/\delta t$

δφ/δt

---

49 FIND THE VOLUME ELEMENT dV IN (a) PARABOLOIDAL (b) ELLIPTIC
    CYLINDRICAL, AND (c) BIPOLAR COORDINATES.

(a) dV = (h1 du1 e1).(h2 du2 e2) x (h3 du3 e3) = h1h2h3 du1 du2 du3
     paraboloid coordinates (u, v, φ)
     x = uv cosφ   y = uv sinφ   x = (u^2 - v^2)/2
     hu = hv = (u^2 + v^2)^.5   hφ = uv

(b) ELLIPTIC CYLINDRICAL COORDINATES.

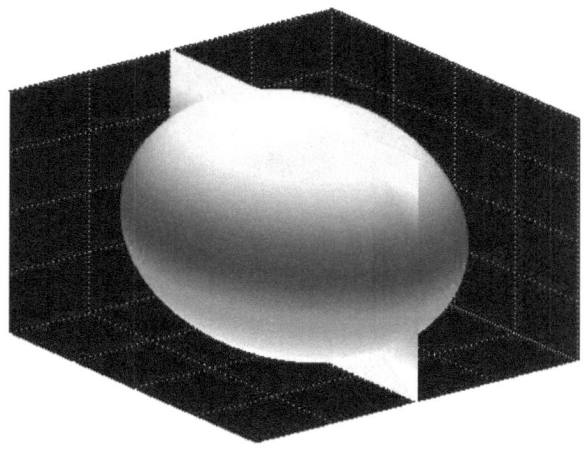

     X = acoshu cosv, y = asinhu sinv, z = z
     hu = hv = a((sin h u)^2 + (sin v)^2)^.5, hz = 1
     dV = hih2h3 du1 du2 du3 = (a(sinh u)^2 + (sin v)^2)^2 du1 du2 du3
     dV = (a(sinh u)^2 + (sin v)^2)^2 du1 du2 du3

(c) BIPOLAR COORDINATES.
     x = (a sinh v)/(cos h v - cos u), y = (a sin u)/(cos hv cos u), z = z
     hu = hv = a/(cos hv – cos u), hz = 1
     dV = h1h2h3 du1 du2 du3 = (a/(cos hv – cos u))^2(1) du1 du2 du3

52 FIND THE ELEMENTS OF AREA OF A VOLUME ELEMENT IN (a) CYLINDRICAL. (B) SPHERICAL, AND (c) PARABOIDAL COORDINATES.

(a)   cylindrical coordinates (ρ, φ, z)

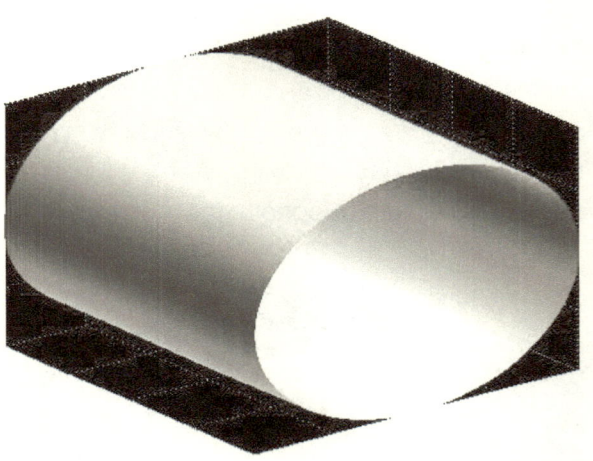

$X = \rho \cos\varphi$, $y = \rho\sin \varphi$, $Z = z$, $dx = - \rho\sin \varphi$, $dy = \rho \cos\varphi$

$h\rho = 1$, $h = \rho$, $hz = 1$

$ds^2 = h1^2 du1^2 + h2^2 du2^2 + h3^2 du3^2$

$ds^2 = 1(- \rho\sin \varphi)^2 + \rho(\rho \cos\varphi)^2 + (1)^2$

$ds^2 = \rho^2((\sin \varphi)^2 + (\rho\cos \varphi)^2) + 1$

$dV = h1\ h2\ h3\ du1\ du2\ du3$, $dA = h1\ h2\ du1\ du2$, $dA = h1\ h3\ du1\ du3$

$dA = h2h3\ du2\ du3$,

$dA = \rho\ d\rho d\varphi$, $dA = d\rho dz$, $dA = \rho\ d\varphi dz$

(d)   spherical coordinates.

$X = r \sin\theta\cos\varphi$, $y = r \sin\theta\sin\varphi$, $z = r\cos\theta$

$hv = 1$, $h\theta = r$, $h\varphi = r\sin\theta$

$dA = h1h2\ du1\ du2$, $dA = hih3\ du1\ du3$, $dA = h2h3\ du2du3$

$dA = (1)(r)\ drd\theta$ $dA = (1)(r\sin\theta)drd\varphi$ $dA = (r)^2(\sin\theta)d\theta d\varphi$

(e)   Paraboloid coordinates. $X = (u^2 – v^2)/2$, $y = uv$, $z = z$

hu = hv = (u^2 + v^2)^.5, hz = 1

dA = h1 h2 du1du2,　　　dA = h1h3 du1 du3,　　　dA = h2h3du1 du3

dA = (u^2 + v^2)du dv,　　dA = uv((u^2 + v^2)^.5)dudz,

dA = uv((u^2 + v^2)^.5)dvdz

HIPERBOLOID

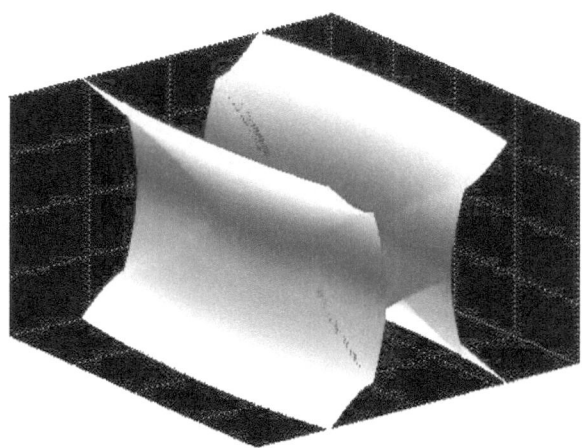

HIPERBOLOIDE

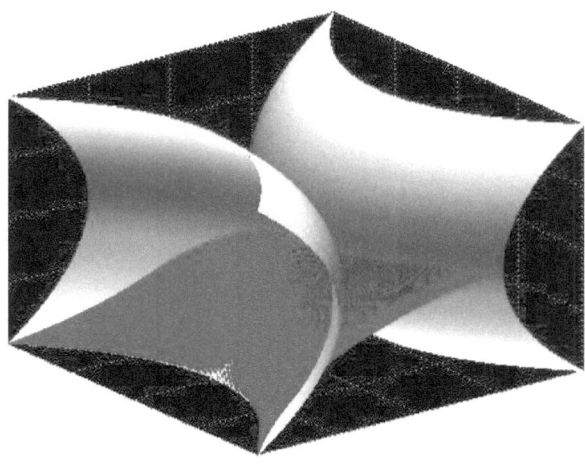

BIBLIOGRAPHY

VECTORIAL ANALYSIS   SCHAUM'S

CALCULO INFINITESIMAL Y GEOMETRIA ANALITICA   GEORGE THOMAS

TRIGONOMETRIA PLANA Y ESFERICA   WEBSTER WELLS S.B.

GEOMETRIA PLANA Y DEL ESPACIO   JORGE WENTWORTH

GEOMETRIA ANALITICA   H.B. PHILLIPS

THE STUDENT EDITION OF MATLAB THE LANGUAGE OF THECNICAL
COMPUTING

www.ingramcontent.com/pod-product-compliance
Lightning Source LLC
Chambersburg PA
CBHW022122170526
45157CB00004B/1721